Santiago Mata: 1917 American Pandemic

Santiago Mata: *1917 American Pandemic. The Origin of the* Spanish *Flu*

A century after its appearance, the pandemic that killed in 1918 tens of millions of people is still unjustly called Spanish Flu. To end this unfair allocation, it is necessary to show how this pandemic emerged in the United States. And to discover that the most deadly virus in History did not begin its expansion in 1918, but in 1917. Finally, it is important to know why the United States Army played a key role in infecting the world with the American Flu, and how this role was hidden by giving the adjective *Spanish* to the influenza.

Web: http://centroeu.com

Email: amanecer@centroeu.com

First published in Spanish (2017):

Cómo el Ejército Americano contagió al mundo la Gripe Española.

To Ángel

Santiago Mata

1917 American Pandemic

The Origin of the *Spanish* Flu

Index

Presentation	9
The 1918 Flu... started in 1917	11
1. Camp Greene: 20 influenza deaths in December 1917	16
2. Camp Dodge: 8 killed by flu in December 1917	23
3. Camp Pike: 12 dead in October 1917	24
4. Camp Bowie: 172 deaths from pneumonia in November	26
5. Camp MacArthur: 7 dead in December 1917	27
6. Camp Beauregard: 50 dead in November 1917	28
7. Camp Cody: From 10 to 30 dead (November-December)	30
8. Camp Grant: Four dead in January 1918	31
9. Camp Custer: seven dead in January 1918	32
10. Camp Sherman: Five dead in December and 8 in January	32
11. Camp Taylor: Four dead in November and 15 in December	33
12. Camp Dix: a dead man in January 1918	34
13. Camp Devens: Four dead in January	35
14. Camp Lewis: Two dead in November	36
Other traces of the *herald wave*	39
The "first" wave of flu in North America	63
Transfer to Europe: Spain, the scapegoat	77
The mass murderer: The second wave	117
Protective sequestration, the West and Overseas	155
The third wave	173
The replicants won over winners and losers	179
Comparison with the covid-19 pandemic	183
Bibliography	185
Acknowledgment	197

Santiago Mata: 1917 American Pandemic

Presentation

A century after its appearance, the pandemic that killed in 1918 tens of millions of people is still unjustly called Spanish Flu. To end this unfair allocation, it is necessary to show how this pandemic emerged in the United States. And to discover that the most deadly virus in History did not begin its expansion in 1918, but in 1917. Finally, it is important to know why the United States Army played a key role in infecting the world with the American Flu, and how this role was hidden by giving the adjective *Spanish* to the influenza.

The 1918 Flu... started in 1917

In 1918, a rural doctor in a small county in the American state of Kansas found a particularly virulent outbreak of influenza. The doctor's name was Loring Miner and the county was Haskell. The locality in which he lived, called Santa Fe, does not even exist today. It has been supposed that there arose the deadliest epidemic that humanity has known.

Haskell forms a perfect quadrate whose sides measure 38.70 kilometers on the Kansas map. It has therefore almost 1,500 square kilometers: somewhat less than the Spanish island of Gran Canaria. Seen from a satellite, one would say that there is nothing relevant in it. Only circles that indicate the use of pivots to irrigate crops. There are not even large properties, since these circles are almost all the same size: 400 meters radius (and therefore half a square kilometer in area); although there are some whose radius is exactly double (and therefore its extension two square kilometers).

No city stands out on the map of Haskell, and only one road, the U.S. Route 56, crosses the county from the southwest to the center-east, passing through two locations: Satanta and Sublette. In Satanta a railway line unfolds and has a deviation towards the northwest, to enter the county of Grant, whose capital is Ulysses, evidencing that both were founded (in 1873) in honor to the 18th president of the United States, General Ulysses S. Grant.

Almost in the southwestern corner of Kansas, Haskell is even younger than its neighboring county, as it was instituted on July 1, 1887, in honor of Dudley C. Haskell, congressman for Kansas born in Lawrence - on the other side of the state, more than 520 kilometers in the direction of Kansas City- and died on December 16, 1883. The world could have continued to turn without taking into account the existence of this plateau at 890 meters of altitude, without shield or flag, inhabited today by 4,300 people -in addition there are two counties of the same name in Oklahoma and Texas-, except for the fact that it has been held responsible for the origin of the 1918 Flu.

Loring Vinton Miner is the first doctor of known name who registered the outbreak of the flu epidemic of avian origin caused by the H1N1 virus. He was born in 1860 in the county of Athens, state of Ohio. In the 1850 census, the municipality of Ames (created in 1802) records as the family number 1,867

Santiago Mata: 1917 American Pandemic

the one composed by the *laborer* Nathan H. Morin (aged 19, born in 1831 in Ohio), his wife Julia (23 years old and therefore born about 1829) and his mother Mary A. (52 years old and born in New York circa 1798). The house was valued at $ 200.

Twenty years later, when the future doctor was already ten years old, the Miners (the sheet of the census of July 19, 1870 called them Moner) lived in the house number 181 of the municipality of Dover (constituted in 1811) , 13 kilometers west of Amesville (the post office on which their residence depended). Nathan, 38, was a farmer and Julia was 43 years old. Their eldest son, Horace L., 16, went to school, as did Irving (14 years), Loring and the youngest, John, 8 years old.

After ten more years, according to the census of June 18, 1880, the Mi-ner were still living in Dover (in the house number 209). The age of the father, Nathan, was the correct one (49 years old), but that of the mother, Julia (who must have been around 53), was down to 41. Horace L. was 26 years old, the next brother, 24 years old, is called here Orwin W., and the youngest John C., was 18 years old.

After studying medicine at the University of Ohio (founded in 1804 in Athens as the oldest in the American West, with the name of American Western University), Loring Miner, apparently a member of the Democratic Party and well received in political circles, is defined by Barry (p.92) as "an unusual man, a classicist enamored of ancient Greece", who toured in 1885 the 1,650 kilometers that in a straight line separate his county in Ohio and Haskell, crossing his own home state , plus those of Indiana, Illinois, Missouri and Kansas to settle in Santa Fe as a doctor who did not just had a consultation, but also a scientific laboratory, a drugstore and a grocery shop.

In 1890, Dr. Miner married a woman ten years younger than he, Lorina, according to the name with which she registered twenty years later in the census of May 2, 1910, when the Miners were the 182nd family of Santa Fe. Loring was then 50 years old, his wife 39, and of their three children two had survived until then: Oliver, 15, and Eugene, nine.

The 1918 flu... started in 1917

The deceased daughter of the Miners, the first-born, was Leah J., born on December 11, 1890 in Eminence, today a ghost town in Finney County (north of Haskell), where the post office opened in 1887 and closed in 1942. Leah died of typhoid fever on August 5, 1907 in Garden City, capital of Finney, in whose cemetery she shares a grave with her parents.

On April 13, 1911 were born the twins Lorene and Loring, nicknamed Sis and Pill. In 1912, the photographer Francis Marion (1866-1936) portrayed them in a field of kafir, a South American corn variety that was all the rage in Kansas after being introduced by the US government - belonging to their father: the two children, healthy looking and blond, wear a light apron and dark stockings.

The census of January 2, 1920 states that Dr. Miner is 59 years old and his wife, already registered as Lorena, 48, while their son Eugene is 19 and is a seller in a drugstore. The twins are 8 years old. The doctor was from Ohio like his parents, while his wife was from Kentucky, and her parents were from Indiana.

In 1930, the twin Loring appears as a sworn (but not active) member of the Kappa Sigma fraternity, founded in 1869 at the University of Virginia, which had a branch for Kansas located at the 1537 house on Tennessee Street in Lawrence, where today the Theta Epsilon chapter of the Pi Kappa Phi fraternity is housed.

The son's student adventure ends when his father dies in February 1935 (he was buried on the 10th). Loring junior began to work as a postman in 1936, although he was not formally appointed until 1939. The census of April 8, 1940 will specify that at 28 years of age Loring had studied up to the third year of high school, was the postmaster and worked 48 hours a week, earning $ 1,800. The house in which he lived with his 69-year-old mother was valued at $ 1,500. Lorena B. Miner was buried with her husband and daughter in Garden City on October 6, 1946. The youngest of the sons and homonym of the doctor Loring Miner died on September 27, 1978, according to the registry of army veterans.

Santiago Mata: 1917 American Pandemic

If we know little about Dr. Miner, there is even less information about the flu outbreak in Haskell in 1918. Not only the doctor has left no memories, nor has there been any trace of the laboratory in which he was able to study the epidemic. The city of Santa Fe itself has disappeared. There was founded in 1887 the first school of the county and in 1897 the first Methodist church.

The town reached 1,800 inhabitants. Its decline began in 1913, when the railway that paradoxically bore the name of Atchison, Topeka & Santa Fe ignored it when drawing the line from Dodge City (Kansas) to Elkhart (Texas), forcing the inhabitants to emigrate to two new localities: Sublette or Satanta. Most of the houses, as The *Hutchinson News* reported on September 10, 1912, would be moved - literally, by mules or wagons - to Sublette, six miles to the south.

On July 25, 1918, the newspaper *Santa Fe Monitor*, created in 1888, closed its offices to move to Sublette. On May 16, 1919, with only 75 inhabitants, Santa Fe lost the vote for the county's capital, although it appealed the result, resolved in December 1920 by the Supreme Court of Kansas in favor of Sublette. There are no visible remains of Santa Fe.

Given the virulence of the registered cases, and although the flu was not one of the notifiable diseases, Dr. Miner asked for advice and help to the health service (US Public Health Service), which did not help him in any way, and just published in the bulletin of April 5, as notified on March 30, the existence of "18 cases of severe influenza, which resulted in three deaths."

This notice is the first document that shows without a doubt that the 1918 influenza pandemic emerged in the United States, even though the notes published in the *Santa Fe Monitor* on influenza are earlier (February). Even more important, in 1929 a book was published that documents the existence of the pandemic in 1917.

The book referred to is Volume XII of the history of the US military health in the First World War -*The Medical Department of the United States Army in the World War,* written by George R. Callender and James F. Coupal under

The 1918 flu... started in 1917

the direction of Charles Lynch-, dedicated to respiratory diseases and damage caused by gas (*Pathology of the Acute Respiratory Diseases and Gas Gangrene following War Wounds*).

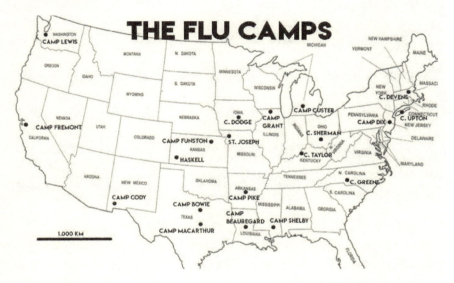

This work studies the diseases that occurred in 16 military camps for the training of recruits for the European war. The data of all of them reflects the existence, in the autumn and winter of 1917-1918, of an epidemic of "acute respiratory diseases" of singular virulence. Actually, the book does not do more than collect the data sent at the time (1917 and 1918) by the doctors of those camps, but when dealing with them after more than a decade, its authors feel with permission to call things almost by name, instead of hiding reality, as was done until then.

Santiago Mata: 1917 American Pandemic

1. Camp Greene: 20 influenza deaths in December 1917

Beginning with Camp Greene (p.88), military training camp outside of Charlotte (North Carolina, more than 1,100 miles southeast of Haskell County), medical reports sent to the inspector general are mentioned. military health, according to which in December 1917 there were 565 cases of influenza, as a result of which 20 patients died, which implies a mortality rate among patients of 3.5%, which could not be so high if It will be a simple seasonal flu.

We are facing the first notification of a new type of influenza with high mortality, although nothing is said about the morbidity, since the population of Camp Greene is not specified:

> The acute respiratory diseases following mobilization at this camp, aside from the high incidence of measles, were due to influenza and common respiratory diseases. 565 cases were diagnosed influenza in the month of December, 1917. The deaths for December were 24 from complications of measles and 20 from primary pneumonias, the latter undoubtedly representing the terminations of unfavorable character in the influenza outbreak. Another outbreak of respiratory

The 1918 flu... started in 1917

diseases, during a time when measles was dropping in incidence rate, occurred with a peak in
April, the majority of the deaths being charged to
primary lobar pneumonia.

According to the *Mortality Statistics 1916* published in 1918 by Sam L. Rog-
ers, Director of the Bureau of the Census (US Department of Commerce) "influenza developed in epidemic form in 1916 and caused 18,886 deaths,
which is equivalent to a rate of 26.4 per 100,000 population. This is an increase of 65% over the rate for 1915 (16), which was 75% higher than that
for 1914 (9.1). With the exception of the rate in 1901 (32.2) the rate for this
cause in 1916 was the highest shown in the annual mortality reports, the first
of which was published in 1900. " (p. 38). The rates for Kansas risen between
1914 and 1916 from 9.5 per 1,000 to 25.6 and 39.3, while those for North
Carolina went from 18.6 per thousand to 19.1 and 26.4.

The population of Camp Greene, where the 3rd and 4th infantry divisions of
the US Army were trained, is estimated at 40,000 people in December 1917.
According to the third chapter of the book by Mitchell and Perzel, the first
unit to arrive , on September 6, 1917, was the company L from Camp Sevier
(Greenville, South Carolina). Then came 10 companies from New England,
with 940 officers and 29,599 soldiers. The population reached 60,000 soldiers, but in the first three months it did not exceed 14,000, when according
to these authors only four soldiers died in Camp Greene.

To obtain the mortality rate for influenza in December 1917, if we assign to
this disease only the 20 deaths that according to the Volume XII of *The Medical Department of the United States Army in the World War (Pathology of
the Acute Respiratory Diseases...)* were due to pneumonias derived from flu
(discounting the 24 caused by complications of measles), we would have a
rate of 33.3 per 100,000 inhabitants if they were 60,000, 50 if they were
40,000 inhabitants and 66.7 if they were only 30,000 (although the 10 companies of New England already exceeded that number).

The figure of 33 deaths per 100,000 inhabitants would be only slightly higher
than normal at that time in North Carolina (where the death rate in 1916 was

Santiago Mata: 1917 American Pandemic

26.4, exactly the same as in the United States). But, as L. Simonsen explains in his 1999 article (p.6), "during the inter-pandemic years, over 90% of influenza-related deaths occurred among persons 65 years of age or older; consequently, the incidence of influenza-related mortality was 100-fold higher for these age groups than among younger persons. "

The mortality rate for influenza among young people should therefore be around 1916 less than 0.3 per 100,000. Values above 30 or even 60 in the fall of 1917 at Camp Greene were between 100 and 200 times what would correspond to a seasonal flu. It was a different flu, whose fatal cases fell on the young. It was the flu caused by the H1N1 virus that in 1918 would become a pandemic.

Regarding mortality over the percentage of patients (3.54%), it is not easy to compare with data from the time, but, despite the difference of almost a century, it can be compared with the modern data provided in the United States by the Centers for Disease Control and Prevention (CDC) with data from patients with influenza and deaths from pneumonia plus influenza:

Season	Sicks	Deaths N+G	%
2010-2011	5.039.277	3.434	0,068
2011-2012	1.981.571	1.227	0,062
2012-2013	5.628.332	1.823	0,032
2013-2014	6.683.929	3.840	0,057
2014-2015	1.606.813	1.419	0,088
2015-2016	5.083.498	2.882	0,057
Total	**26.023.420**	**14.625**	**0,056**

The mortality of 3.54% in Camp Greene multiplies by 63 the modern of 0.056%, and if we keep in mind the fact that 90% of those deaths are people aged 65 or older, it is clear that what happened at Camp Greene had nothing to do with a seasonal flu.

The 1918 flu... started in 1917

What is strange in this camp is the apparent low morbidity, since if there were 565 cases of flu in a population of 40,000 soldiers, only 1.4% of the inhabitants were ill. Anyway, with 1,400 cases per 100,000 inhabitants, it is undoubtedly an epidemic, since these are declared after 15 confirmed cases. But it is far below the morbidity that will be typical for the pandemic, so we can only suspect that there were many patients with influenza who did not get to the camp hospital.

The statistics for Camp Greene is completed (p.90) with data from an article published in December 1918 by Captain Herman Elwyn in the *Southern Medical Journal*, reporting 427 cases of lobar pneumonia from mid-December 1917 to April 1918, with complications (mostly emphysema), in which death (p.91) occurred within a period of one or two days, with characteristics that will be typical of pandemic influenza in its deadliest version:

> The face assumed a grayish color, was pinched and anxious, the pupils moderately dilated. The temperature rose promptly to 103 or 104°. The pulse was rapid, usually about 140 to 160. Breathing was shallow and rapid and the breath sounds on the affected side were diminished or absent. The prostration was extreme. Within four to six hours after the onset, fluid could be obtained on aspiration over the base of the lung. The fluid increased rapidly and in 12 to 18 hours could be percussed above the angle of the scapula. Dyspnea and prostration were present; pulse rate increased to a marked degree; the pupils became completely dilated; the patient sweated profusely; and death often resulted within 24 to 48 hours.

According to Elwyn, 34 cases of lobular pneumonia appeared as a complication of measles, of which 17 died, and 393 were "primary," of which 68 died (85 deaths in total). The article avoids talking about flu, because it was published during the second wave of the pandemic, but says that "fever" lasted between five and ten days and was cured when the pulse remained between

Santiago Mata: 1917 American Pandemic

110 and 120, but that a rise above 140 "bespoke grave prognosis".

Elwyn specifies the beginning of the epidemic by saying that cases of pneumonia were "comparatively few 'til about the middle of December." The medical captain also gives an indication that a more severe wave of the epidemic began in February, when he said:

> During December the cases began to show a tendency to involve more than one lobe. During February they seemed to assume a somewhat different character and were more severe. Besides the multiple lobe involvement, the patients were usually pale, showed no herpes labialis, had an increased pulse rate out of proportion to the ordinary cases, and in general they seemed to be in a more or less typhoid state.

Elwyn's article was "approved for publication" by the Surgeon General of the U.S. Army, and symptomatically "prepared for *Section on Medicine, Southern Medical Association*, 12th Annual Meeting, Nov. 11-14, 1918, postponed one year on account of influenza epidemic". The author assures that many cases looked like meningitis, "and the necessity of confirming or disproving this promptly led us to tap the spinal canal, but in a few we were able, after making them cough repeatedly, to obtain rusty sputum and thus save ourselves and the patient the spinal puncture".

Avoiding to mention the flu, Elwyn excuse the diagnostic failures adducing the strangeness of the symptoms:

> That mistake in diagnosis in such conditions is not made only by the novice can be seen from the following. A very distinguished surgeon visiting Camp Greene happened to see a case at one of the regiments and sent the man to the hospital with the diagnosis of acute appendicitis. The Officer of the Day who received the case obtained rusty sputum on making the patient cough and

sent the man to the pneumonia ward. Next day all
the signs of pneumonia appeared.

The fact that the disease was not pneumonia could be inferred from what
Elwyn says when talking about the most serious cases that began to appear
in February, when it also took longer to study the disease:

During February we had a number of cases which
were characterized by a rapid pulse, pale color instead of the flush which one notices ordinarily on
the faces of pneumonia patients, a somewhat typhoid state and a tendency to involvement of
both lungs. A good many of these recovered, but
their course was usually long.

Once again, there is a temporal reference when talking about complications,
specifically the accumulation of pus (empyema), where the data collected
would have ceased on March 1, with 14 cases derived from measles and subsequent bronchitic and lobular pneumonia (10 dead, 71.4% of the affected);
29 cases after measles without pneumonia (18 deaths, 62%); 14 without
measles or pneumonia (4 deaths, 28.5%) and 45 among the 393 cases of primary lobar pneumonia (21 deaths, 45%).

It is precisely talking about the most serious empyemas when Elwyn makes
the aforementioned description about the symptoms of the disease, which
is preceded by an emotional declaration of surprise that, knowing what he
was really facing, hardly leaves any doubt that he was in the presence of the
pandemic flu (and, possibly, that to that disease should be attributed the
deaths that Elwyn considers complications of measles):

These often appeared after the pneumonia had
subsided; they appeared also in the course of
measles, bronchitis and in a few cases secondary
to a tonsillitis or without any apparent cause.
They presented a clinical picture totally unlike
any other cases of empyema that the writer ever
saw.

> The onset was with a sudden sharp pain in the side or in the abdomen. The most striking cases were those after measles in which normal convalescence had seemed established and the patients were up and about the wards with normal temperatures. The pain was intense, the most intense I have ever seen, and in at least one case where the pain was in the abdomen we suspected acute peritonitis in addition, which, however, was not found at the autopsy.

The attempt to pass off pandemic influenza as pneumonia could have been intentional or naive. But the fact that measles also appears, and that as we saw in Camp Greene more deaths are assigned to this disease than to influenza, raises the question of whether this disease could also hide the flu, because the language of doctors tries to relate measles and pneumonia as if the second could be derived from the first, avoiding at all costs talking about flu. It is symptomatic that this is what is done in the introduction to the Volume XII of the history of *The Medical Department (Pathology of the Acute Respiratory Diseases...)* making a judgment or at least a model to fit what happened in all military camps in the United States (page 7):

> Measles in epidemic form made its appearance in our newly mobilized troops in the fall of 1917 and as the epidemic progressed the incidence of pneumonia as a complication increased just as it has in former wars. In most of the camps, in addition to measles, there was a high incidence of acute respiratory disease independent of measles, which assumed epidemic proportions and was accompanied by a large number of cases of pneumonia. At first the pneumonia corresponded with that seen in civil life, and this was particularly true in the northern camps. Mortality increased, however, in cases following the acute respiratory diseases and following measles, and it became evident that organisms of unusual virulence were

present. This epidemic condition waned about the beginning of 1918.

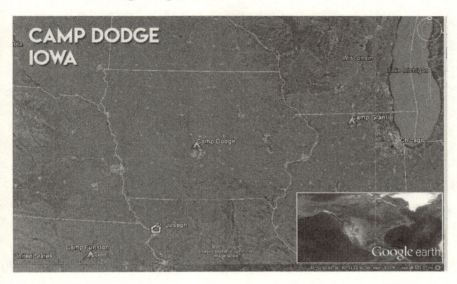

2. Camp Dodge: 8 killed by flu in December 1917

If we turn our attention to the center-north of the United States, at Camp Dodge (located in Johnston, Iowa, 475 miles northeast of Haskell in a straight line), the Volume XII of *The Medical Department* (page 62) recognizes that shortly after the first contingents of soldiers arrived in that field (September 1917) "measles, together with other acute respiratory diseases, made its appearance in epidemic proportions. A peak of incidence is shown in December at which time the case fatality rate for all respiratory diseases was 0.7%. 265 cases were diagnosed influenza in this month, though the deaths were attributed, for the most part, to primary pneumonia."

Eight deaths are a significant 3% in mortality if, as seems to be suggested, they should be compared with the 265 cases of influenza. Anyway, of the thousand admissions in December, a little less than 200 were for measles, the same amount as in the peak of March, when more than a thousand admissions caused more than 30 deaths, which in April would increased to 40, although the number of admissions dropped to just over a thousand.

3. Camp Pike: 12 dead in October 1917

For Camp Pike (today Camp Robinson, in North Little Rock, Arkansas, 520 miles southeast of Haskell, almost in the geographical longitude of Camp Dodge and almost in the latitude of Camp Greene), the Volume XII of *The Medical Department* (p.125) hardly interprets the data, despite the fact that the epidemic had a great incidence (or maybe because of this): the admissions due to acute respiratory diseases were 700 in October 1917 (more than 500 due to measles), with a dozen deaths; Doubling in November (around 1,400, with more than 800 cases of measles and 30 deaths) and repeating the data in December (with measurable decrease of measles down to 550 admissions), when a peak of almost 60 deaths was reached.

In January 1918, the admissions to the Camp Pike hospital were still in excess of 1,000, with little more than 100 for measles and 20 deaths. The minimum point of the epidemic - it can not be said that it disappeared - occurred in February with more than 500 new admissions (only 20 due to measles) and three deaths.

Although the report refuses to assign deaths to the flu, and also when evaluating cases together from the autumn of 1917 to the spring of 1918, it does not allow to discriminate symptoms, it mentions (without giving the date)

The 1918 flu... started in 1917

fulminant cases, like a patient who entered at 7:00 pm one day and was dead at 6:00 the next morning (p. 127):

> On examination, he presented practically no signs of pneumonia, except a few moist râles which were distributed over both lungs and not confined or isolated to any particular part of the lung. He did, however, show the following signs of meningitis: Headache, depression, hyperesthesia and stiff neck. There was no rash, no Kernig sign, no Babinski reflex, no Oppenheim sign and no Gordon reflex. The diagnosis of bronchopneumonia was not made before death on account of the trivial findings. The post-mortem examination disclosed a small number of bronchopneumonia patches in the left lung and in the lower lobe of the right lung. There were no signs of meningitis. There was an acute and well-marked lymphadenitis of the bronchial lymph nodes. There was a cloudy swelling of the liver and kidney and an acute splenitis.

This report on Camp Pike includes the only mention of Camp Funston - the military camp closest to Haskell – in the Volume XII of *The Medical Department* (p.131), just to establish the relationship between bronchitis and influenza, trying to base it on the presence of the bacillus that was erroneously considered responsible for the flu at the time:

> In attempting to establish the relation of *Bacillus influenzae* to influenza and its complications, it was borne in mind that at Camp Funston, *B. influenzae* was found in the mouths of 35.1% of all healthy men examined and was present, in the absence of an epidemic of influenza, in the sputum of a very large proportion of those suffering with bronchitis. Observations at Camp Pike showed that the organism was invariably present

in the upper respiratory passages of patients with influenza.

4. Camp Bowie: 172 deaths from pneumonia in November

Camp Bowie (on the heights of Arlintong, Fort Worth, Tarrant County, Texas, 390 miles south-southeast of Haskell), recorded from September 24, 1917 to January 1, 1918 a measles epidemic (p.22-23) with 3,624 cases, in addition to 973 cases of pneumonia (only in 363 cases preceded by measles), which caused 237 deaths.

Although the report avoids talking about influenza in Camp Bowie, suddenly we find (p. 24) that, precisely after saying that in November 1917 there were 127 deaths from pneumonia after measles and 45 from primary pneumonia, "it is probable that influenza in epidemic proportions preceded many if not most of the cases of primary pneumonia and may well have been a coincident infection with measles in many cases. The case fatality for measles was 4.01%, while that for the cases of measles-pneumonia was 39.18%. "

The 1918 flu... started in 1917

5. Camp MacArthur: 7 dead in December 1917

For Camp MacArthur (Waco, Texas, 80 miles south of Camp Bowie), the Volume XII of *The Medical Department* recognizes that in November 1917 there was an epidemic of acute respiratory diseases, "being t first, however, relatively little pneumonia and few fatalities," that reached a peak in January 1918 "accompanied by a relatively high fatality rate". The report seems to want to make the measles disease responsible for the epidemic, because it showed that month "the highest rate for the period of the war", but it recognizes to the 199 cases of influenza diagnosed that month the character of "epidemic rise". In April 1918, with a "very low" mortality rate, it identifies a peak for which "influenza and common respiratory diseases were responsible. "

If we follow the figures presented in this report (p.108), in November 1917 admissions were more than 350, with almost 200 cases of measles and two deaths, data that in December amounted to more than 450 admissions, less than 250 for measles and seven deaths, and in January for almost 700 admissions, measles stagnated in 250 and the deaths rise to almost 25. In February, the admissions fell to 400, the measles cases to 60 and the deaths to six, the epidemic situation remained in March with 450 admissions (less than 25 for measles) and four deaths.

Santiago Mata: 1917 American Pandemic

Three of the deaths cataloged at Camp MacArthur in January 1918 as pneumonia were known by the press at the time of the transfer of the 32nd division from that camp to Hoboken (the port of New Jersey, on the banks of the Hudson River, in front of New York) . Terri Jo Ryan, journalist who reviewed the obituary of the *Waco Daily Times Herald*, pointed them out in the *Waco Tribune* on December 29, 2007 as the beginning of the 1918 pandemic in that camp, thus prior to the outbreak of the Haskell epidemic:

> Peter Mauseth, 24, a private in Company F, 337th Infantry, died New Year's Day 1918 in the base hospital at Camp MacArthur of pneumonia. "He was taken off the train ill," the notice said. Percival Risher of Wisconsin, 19, a private in the headquarters company of the 119th Machine Gun Battalion, died Jan. 12 in the base hospital at Camp MacArthur of lobar pneumonia — just hours before his mother arrived by train to tend to him. Elmer E. Ranck of New Jersey, a private in the 41st Recruit Squadron, 3rd Provisional Regiment, died Jan. 16 of pneumonia at the base hospital.

6. Camp Beauregard: 50 dead in November 1917

In Camp Beauregard (Louisiana, 650 miles southeast of Haskell and 240 south

28

The 1918 flu... started in 1917

of Camp Pike) the report talks of "a definite epidemic outbreak" of influenza during the months of March and April of 1918, but as if it was a sandwich between two most serious measles outbreaks (p.11):

> Measles was a serious factor at this camp and was responsible for a considerable number of deaths, showing a sharp increase at the time of the original mobilization, and just before and just after the influenza epidemic.

The graph that the book attached to that camp assigns to measles more than a thousand cases in November 1917, when there were 50 deaths, supposedly assigned to that disease. In December 1917, there are still 600 cases of serious respiratory diseases, but only 90 are measles, and there are 30 deaths.

In January 1918, with only four cases of measles, there are still more than 400 cases of acute respiratory diseases, and eight deaths, which in February are reduced respectively to 260 cases and two deaths, and in March - with a couple of cases of measles as in the previous month, or April and May- will be 200 and zero deaths. In April the epidemic returns (already recognized as influenza) with 550 cases of acute respiratory diseases and four deaths, while in May there are 350 cases and three deaths.

Santiago Mata: 1917 American Pandemic

7. Camp Cody: From 10 to 30 dead (November-December)

In Camp Cody (New Mexico, 535 miles southwest of Haskell) the report talks of an influenza epidemic in November, but the deaths were "were charged very largely to the primary pneumonias" and autopsies of that period are not preserved (p. 31):

> The epidemic of respiratory diseases, aside from measles, was due to influenza and common respiratory diseses. During the beginning of this outbreak, or in the month of November, the case fatality rate for all respiratory diseases was 0.95%. With the continuation of the epidemic, the streptococcus appeared in considerable numbers and the case fatality rate increased to 1.71% in December.

In any case, it is noteworthy that unlike what was done in Camp Bowie here mortality is not attributed to measles, among other things because the prevalence of this disease was lower. Hospital admissions in November 1917 were below 1,500 and in November they exceeded that figure. In both months the measles cases did not reach 100, while it went from just over 10 dead to almost 30. In January 1918, there were still 650 admissions and 15 deaths, with less than 20 cases of measles.

Although there are no laboratory tests, the peak of hospital admissions at Camp Cody suggests that here too the first wave of pandemic flu was in December 1917. In any case, admissions and deaths continue to decline in February (350 and 7) and March (280 and 6), picking up so little in April (300 admissions and 5 deaths) that it does not seem credible to identify that peak as the first wave of pandemic flu.

The 1918 flu... started in 1917

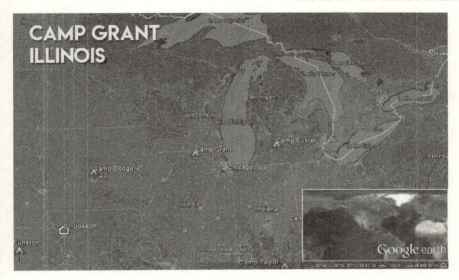

8. Camp Grant: Four dead in January 1918

As we do to the south, we find traces of an early onset of the epidemic in the north of the United States. At Camp Grant (west of Chicago, 700 miles northwest of Haskell and 240 east of Camp Dodge), the Volume XII of *The Medical Department* (page 81) recognizes an early epidemic, not in December of 1917, but in the following month, when speaking of "a relatively low case fatality at the peak of incidence in January, 1918: 0.6% of the total respiratory disease incidence. The deaths were attributed to primary lobar pneumonia. During February, 1918, the incidence of inflammatory respiratory diseases declined, but the case fatality rose sharply to nearly double that of January, or 1.10%. The occurrence of the streptococcus among the cases increased at this time, as did empyema. "

Specifically, if by December 1917 there had been almost 400 admissions for acute respiratory diseases in the Camp Grant hospital (and at least one death), in January 1918 there were 700 admissions (200 for measles) and four deaths, figure which was repeated in February, increasing the rate because admissions had decreased to 350 (only 45 per measles). Admissions fell very little in March (and only 20 were due to measles), when there were three deaths, so the supposed first pandemic wave in April was a rebound in mortality (16) more than in admissions (400).

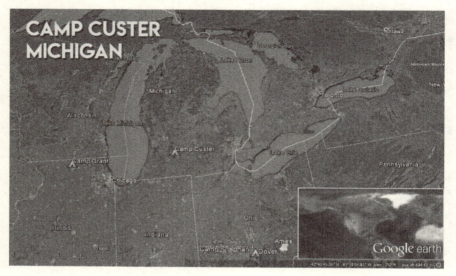

9. Camp Custer: seven dead in January 1918

East of Lake Michigan (and almost 200 miles in a straight line from Camp Grant) was Camp Custer, where the Volume XII of *The Medical Department* recognizes a mortality rate in respiratory diseases that rises from 0.32% of the patients in December 1917 (350 admissions, 100 of them for measles) to 1.15% in January 1918 (600 cases, measles remaining in the 100, and seven dead), which is qualified as normal. The official story does not recognize influenza epidemic until March-April.

10. Camp Sherman: Five dead in December and 8 in January

In Camp Sherman (Ohio, almost in the line from Camp Custer to Camp Greene, 980 miles east of Haskell and 240 southeast of Camp Custer), the Volume XII of *The Medical Department* (page 137) plainly acknowledges that "the case fatality rate for all respiratory diseases reached a high point in January, 1918, at 1.1%, dropping during February to 0.5% and increasing sharply in the month of March, 1918 to 1.16%."

The data of the report reveal that since November 1917 there was an epidemic situation, with more than 450 hospital admissions -less than 20 were measles- and one death, reaching in December almost 600 admissions (55 for measles) and five deaths, in January almost 800 admissions (more than

The 1918 flu... started in 1917

200 for measles) and eight deaths. In February, despite the above said, the epidemic increased with over 800 admissions (one hundred per measles), although the deaths were certainly reduced by half (four).

In this context, although the report once again mentions the peak of March in the chapter dedicated to the first wave of pandemic influenza in 1918, as if it was different from this of 1917 that we can qualify as *herald wave*, continuity seems absolute, except regarding mortality.

11. Camp Taylor: Four dead in November and 15 in December

At Camp Taylor (south of Louisville, Kentucky, 830 miles east of Haskell and 170 southwest of Camp Sherman), the Volume XII of *The Medical Department* recognizes (p.144) that "Influenza, which was present with the measles, increased in December, and this increase was accompanied by an increase in the case fatality rate, ascribed in the records to primary pneumonia, but actually due to pneumonia secondary to influenza, and other respiratory diseases. "

In fact, although the epidemic is less virulent than in other places in its beginning (less than 200 hospital admissions in October, including 50 for measles, and only one death), in November it exceeded 700 admissions (more than 300 for measles), with four dead, and in December reaches a peak that exceeds widely 1,000 admissions (but just over 200 for measles) and causes 15 deaths. In January there were 850 admission (less than 80 for measles) and nine deaths, and after registering a low point in February (500 admissions, only 10 of them for measles, and three deaths), it rebounded in March.

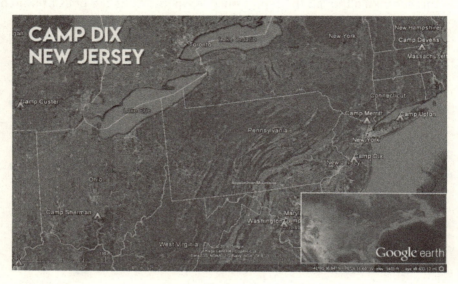

12. Camp Dix: a dead man in January 1918

56 miles south of New York (1,420 east of Haskell and 450 from Camp Sherman), in Camp Dix (New Jersey), the Volume XII of *The Medical Department* admits (p.54) that "measles and common respiratory diseases assumed epidemic proportions shortly after mobilization began in September, 1917. Influenza was diagnosed in considerable numbers throughout the fall of 1917. The sharp rise in respiratory diseases, with peak in March, 1918, was due to an epidemic of influenza, over 1,100 cases being so diagnosed in that month. "

The editors, although having the honesty to admit that those of March 1918 is an influenza epidemic, do not highlight the peak of more than 700 hospital admissions in January 1918 (in October of 1917 they were less than 250, in November, they rise to around 350 and in December they were more than 500), with less than 20 cases of measles and only one death. In the absence of more data, the question remains whether it was a simple seasonal flu or the most virulent type that had already made its appearance, and which would have remitted in February to reappear in March.

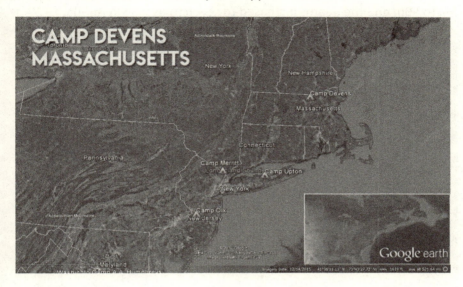

13. Camp Devens: Four dead in January

In the northeastern end of the recently established system of military camps, Camp Devens (in Ayer, Massachusetts, 31 miles from downtown Boston) also records a peak in admissions for serious respiratory diseases in December 1917, suddenly exceeding 600 cases (in November there were just over 200 and in October about 80), with only 75 cases of measles and one death. In January of 1918 it falls below 500 admissions (measles slightly exceeds 100) and there are four deaths. In February, there are almost 400 cases - just over 20 measles - and three deaths.

However, for the editors of the Volume XII of *The Medical Department* (p.42), the only comment worth mentioning is that "whereas the incidence of acute respiratory diseases at this camp, shortly after mobilization, was high, the mortality was relatively low, the case fatality rate for the month of December, 1917, being 0.21%, and for January, 1918, 0.8%. " While this increase does not seem to attract attention, as we will see, the rise to 1.07% in March and 1.97% in April is already considered "sharp", making (p. 44) "quite probable that an influenzal wave occurred at this time. " The truth is that the admissions in March were the same as those in December (600, in April they fell to 500) and the deaths increased to seven (plus nine in April).

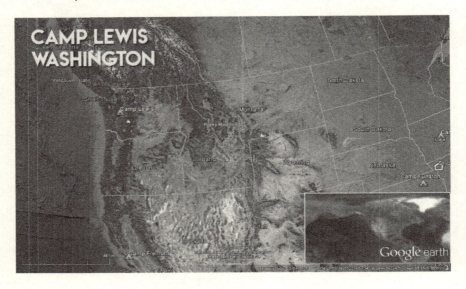

14. Camp Lewis: Two dead in November
In the western States, there is no evidence for an early spread of influenza in Camp Freemont (California, 1,170 miles west of Haskell) and the Volume XII of *The Medical Department* asserts (page 98) that also in Camp Lewis (Washington, south of Seattle, almost 1,300 miles northwest of Haskell) what happened was "contrary to the experience in other camps" because the mortality rate among patients with acute respiratory diseases of 0.3% in November 1917 was lowered to 0, 06% in December, registering the first large increase in February (mortality of 0.56%), a figure that, despite the increase in morbidity in March, fell back to 0.19%.

The comment is somewhat forced, as it is limited to analyzing mortality, giving the impression that there was no epidemic of influenza in the early winter of 1917-1918 in Camp Lewis. The data of November (two deaths) is related to more of 550 hospital admissions, evident epidemic that continues with a slight decrease in admissions in December (and without deaths). January, with only one death and less than 400 admissions may not seem relevant, but the epidemic continues and gets worse, because although morbidity declines in February (around 350 admissions), mortality increases, with two deaths.

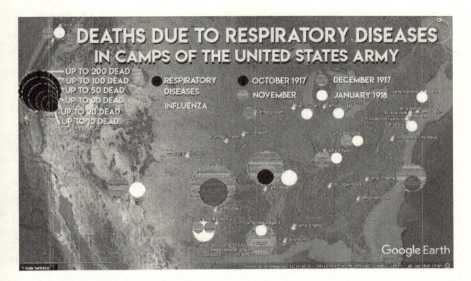

Summarizing the data provided by the North American physicians themselves, it can be very probable that the first wave -called *herald*- of the epidemic of the H1N1 flu virus that would lead to the 1918 pandemic started in October 1917 in the military camps from the center of the United States, with deaths most likely due to the flu since that same month in Camp Pike.

In November there is a mortality rate much higher than that of Camp Pike in Camp Bowie, the first place where over time the influenza-nature of the deaths that documentation assigns to other diseases will be recognized. In addition, the flu epidemic extends southward (Camp Beauregard), where mortality also exceeds that of Camp Pike.

In December, mortality increases in that camp of Arkansas, but decreases in the aforementioned of Louisiana (and probably in Camp Bowie), while this *herald wave* extends eastward (Camp Greene) and with less intensity in other directions.

Since the Volume XII of *The Medical Department* does not collect data for

Santiago Mata: 1917 American Pandemic

this period in Camp Funston (which began to be built in July 1917 and since September began to accommodate units of the 89th Infantry Division), it can not be ruled out that this camp had a role in the transmission of pandemic influenza, but this did not manifest prior as in the camps cited before, so Camp Funston is more likely to be just another link in the chain of epidemic infection. The flu was transmitted from Camp Funston soldiers to civilians of Haskell, although the news published gives the inverse feeling.

The pandemic influenza of unusual mortality that appears in 1917 in the center of the United States was something new for the majority of the doctors of the time, but not for the humanity. In any case, the closest precedent had happened too long ago, so most doctors could not have experienced it.

Since the end of the eighteenth century, there had been five major influenza pandemics: those of 1781-82, 1789-99, 1830-33, 1847-51 and 1889-1894: therefore, a quarter of a century had passed since the last. Between the last two, the Italian doctor Edoardo Perroncito (1847-1936) identified in the north of Italy the existence of this disease among birds, calling the bird flu "typhoid epizootic in the gallinaceous" in an article published in 1878 in the Annals of the Agricultural Academy of Turin.

As for the causes of the flu, between 1901 and 1903 it had been possible to confirm the passage of avian influenza to swine and human, but it was still unknown that its cause was a virus. It was erroneously considered that its cause was a microorganism belonging to the domain or realm of bacteria, ie cells without a nucleus (prokaryotes, unlike the cells of the animal kingdom, plants and fungi, which are called eukaryotes because they have a nucleus) , and in particular of the type that by their shape that combines a bacillus or cane and a sphere or coconut, are called coccobacilli. Specifically, the flu was associated with a coccobacillus then known as *Pfeiffer bacillus* or *Bacillus influenzae* (now *Haemophilus influenzae*) discovered in 1892 by the German bacteriologist Richard Pfeiffer (1858-1945).

Other traces of the *herald wave*

The existence of a *herald wave* prior to the appearance of influenza in Haskell (January) or Camp Funston (March 4) is proven by data from at least 14 military camps (those cited in the previous chapter). But there is evidence of it elsewhere, and in fact I have taken the expression *herald wave* from Brian L. Hoffman, who in an article published in 2011 studied one of these cases.

According to Hoffman, the flu whose symptoms allow us to speak already of the presence of the pandemic virus that would devastate the world in 1918, had already provoked in February 1918 an increase of the mortality by influenza in young people of the city of Saint Joseph, Missouri, bordering Kansas and 360 miles away from Haskell.

Saint Joseph has maintained the 77,000 inhabitants it already had in 1910 for more than a century. Hoffman studied death certificates for influenza and pneumonia, initially perceiving "an increase in pneumonia and influenza mortality starting in the 1915/1916 influenza season. Initially, increased mortality was observed in infants and the elderly. In February 1918, an age-shift typical of pandemic strains of virus was seen, as the burden of mortality shifted to young adults, a characteristic of the 1918 pandemic virus. These results provide one of the first confirmations of the existence of a *herald wave* of influenza activity in the United States prior to the recognized start of the H1N1 pandemic in Spring 1918. "

Assuming that the year 1918 will be recognized as that of the universal flu pandemic, Hoffman points out that the age of the individuals most severely affected is the first difference between that year's flu and the others:

> During prepandemic years, the age-specific pneumonia and influenza (P&I) mortality curve was U-shaped, with relatively high mortality rates in infants and adults over-65 years of age. During the 1918/1919 pandemic year, the mortality curve took on a W-shape, with a peak in mortality in young adults as well as in infants and older adults.

The explanation for excessive mortality "in older adult populations and

Santiago Mata: 1917 American Pandemic

younger populations" would come, according to the studies cited by Hoffman, is that "the robust immune system of a healthy young adult may initiate a *cytokine storm* in response" to the H1N1 virus (which many years later would be identified as the cause of the 1918 pandemic), "resulting in massive cell death in infected tissues."

Although Hoffman provides important data and gives name to the *herald wave*, he does not venture to point out that it could begin in 1917, even if it is clear that he speaks of a winter flu (which, as we have seen from the data of the military camps, actually starts in the fall, as usual with seasonal flu) instead of spring, and he also studies deaths in February, which would allow to assume that, in they were victims of an epidemic, this should have started months ago.

Despite not thoroughly deciphering the chronology of the origin of the flu, Hoffman does explain with simplicity the reason why the pandemic virus affected mainly the strongest individuals and not the weak ones. The immune system is the one that responds to the aggressions of external (pathological) agents, and it is divided into innate (white blood cells or leukocytes) and adaptive, which generates the immunity acquired using antibodies and lymphocytes.

While normal flu viruses weaken organisms, posing greater danger to the less immunologically prepared (children and the elderly), the greatest immunological alarm that causes the strange pandemic virus is not noticeable in the weak, who suffer this flu like any other, but in the strong causes an excessive reaction, which comes to block the airways with the presence of numerous white blood cells, and to drown the individual, so to speak, as if he were strangling himself with his own hands while trying to get rid of a rope.

Hoffman says that in young people there was a "naive" response of a mature immune system to his first exposure to the H1N1 virus, while children would enjoy a "honeymoon period" in which "a general lack of mortality due to serious illness is seen in this age group ". Individuals older than 30 years, at the other extreme, would not register an exaggerated immune reaction to the

Other traces of the herald wave

new type of influenza because they would have been exposed to similar viruses "prior to 1890 and retaining some immunity to that subtype of influenza virus." The elderly also would have "reduced levels of immunity because of underlying illness or senescence."

Specifying the data of his study, Hoffman summarizes that Saint Joseph registered two waves of pandemic activity related to influenza and pneumonia among young individuals: a first mild between February and May 1918, and another with higher mortality between October 1918 and May 1919.

This suggests to Hoffman that the origin of the 1918 pandemic would be prior to the events of Haskell and that therefore "a *herald wave* of pandemic activity sprang from widely separated geographic locations in the United States early in 1918, prior to the generally recognized outbreaks in Kansas. "

Regarding the exact data of the increase of juvenile mortality in 1918 in Saint Joseph, according to Hoffman between February and May there was "highly elevated excess mortality in the under-5 year age groups and elevated excess mortality in the 20-24 year age group", while "mortality in the >65 years age group dropped to 14% of 1915/1916 levels. "

In January 1918, the risk ratio of dying from pneumonia or influenza (P & I) "decreased nearly 65-fold from the previous month" and "was 15-fold lower than the 1915/1916 epidemic period. By March, all of the excess mortality had shifted to the <65 year age-groups, a pattern repeated during the 1918/1919 pandemic season. By spring 1918, the proportion of all excess deaths seen in persons <65 years of age had more than doubled from 1915/1916 levels."

Hoffman concludes:

> The epidemiological evidence in this study supports the hypothesis that a very virulent influenza virus exhibiting characteristics of a pandemic strain was active in the central United States during February 1918, which matches previous accounts of a serious influenza from Haskell

County, Kansas in the same time frame. Taken together with evidence that such a virus was also circulating in New York City in February 1918, it seems likely that precursors to the pandemic strain were widely circulating in the United States prior to the March 1918 outbreak at Fort Riley [Camp Funston], Kansas and subsequent spread through military camps and nearby urban centers. This further contradicts the idea that central Kansas was the point of origin of the 1918 pandemic virus.

About New York - more than 1,450 miles away from Haskell in a straight line - there are two relatively well-known punctual data, the one that there died of pandemic flu the filmmaker Joseph Kaufman on February 1 and on March 27, the five-time Olympic medalist Martin Sheridan. A study by Donald R. Olson and three other authors states that "the timing, magnitude, and age distribution of this mortality shift provide strong evidence that an early wave of the pandemic virus was present in New York City during February-April 1918. "

After presenting the factors combined to obtain the values of what they call *expected deaths at time t* (M_t), with which they evaluate the deaths from pneumonia and influenza (P & I), the authors found two peaks of mortality in January and March 1918 in New York, noting that, as had happened in 1915-16 and 1916-17, the excess mortality in January still affected especially those over 65 years of age.

But "during February-April 1918 and the 1918/1919 season, those ≥65 years old experienced little or no excess mortality, whereas those aged 15-24 and 25-44 years experienced sharply elevated death rates, " leaving young children also safe of the epidemic. Olson and his colleagues conclude that "the ratio of age-specific epidemic deaths shifted abruptly in February 1918 and reached values >20-fold higher in March and April 1918 and in the 1918/1919 season than in the 1915/1916 and 1916/1917 epidemic seasons. The age-specific ratio in the 1919/1920 season was closer to prepandemic levels;

Other traces of the herald wave

however, the total burden of excess deaths among people <45 years old remained extremely high, compared with the interpandemic period."

The summary of that work underlines the importance of New York as a city that at that time harbored more than 5% of the population of the United States, and insists, as Hoffman did, that the presence of the pandemic in the great city in February 1918 would be "inconsistent with the prevailing hypothesis of a spring 1918 Kansas origin."

The first camp that, according to that hypothesis of an influenza pandemic originated in Haskell, would have been infected, was Camp Funston, the second largest instruction field of the US Army, in which more than 56,000 recruits lived in 1,400 buildings on 2,000 acres (8,1 km^2). Today it is part of the military compound of Fort Riley, immediately south of the small town of Ogden, Kansas.

Before the end of March 1918, 1,100 soldiers suffering from influenza were hospitalized in Camp Funston, 20% of them developed pneumonia and 38 died. The epidemic has been assumed, wrongly as we have seen, imported by resident recruits in Haskell County, according to some interpretation of the news about Dr. Miner's patients published in the modest local press and quoted by John M. Barry in his article published in 2004:

> In late January and early February 1918 Miner was suddenly faced with an epidemic of influenza, but an influenza unlike any he had ever seen before. Soon dozens of his patients – the strongest, the healthiest, the most robust people in the county – were being struck down as suddenly as if they had been shot. Then one patient progressed to pneumonia. Then another. And they began to die. The local paper *Santa Fe Monitor*, apparently worried about hurting morale in wartime, initially said little about the deaths but on inside pages in February reported, "Mrs. Eva Van Alstine is sick with pneumonia. Her little son Roy is now able to get up... Ralph Lindeman is still quite sick... Goldie

Santiago Mata: 1917 American Pandemic

> Wolgehagen is working at the Beeman store during her sister Eva's sickness... Homer Moody has been reported quite sick... Mertin, the young son of Ernest Elliot, is sick with pneumonia... Pete Hesser's children are recovering nicely... Ralph McConnell has been quite sick this week (*Santa Fe Monitor*, February 14th, 1918)."

A week later, on February 21, the *Monitor* stated that "almost everyone in the country has the flu or pneumonia," and on February 28, it gave information about visits that could be vectors of the epidemic to or from Camp Funston: "Dean Nilson surprised his friends by coming home from Camp Funston with a five-day permit." Ernest Elliot, whose son as we saw had pneumonia, went to Camp Funston to visit his brother. The same February 28 marched towards the camp John Bottom (a person exposed to the virus like all the inhabitants of the county), on whom the *Monitor* said: "We predict that John will be an ideal soldier".

Barry believes that from Camp Funston influenza spread to other military camps, as the chain would show: on March 18 the first cases of flu are recorded in Camp Forrest and Camp Greenleaf, two of the camps (along with a third called Camp McLean), created in Fort Oglethorpe (Georgia), almost 700 miles southeast of Camp Funston. However, as we have seen, the epidemic could reach Camp Funston months before appearing in Haskell, because it was already present, in its pandemic form, in at least 14 military camps.

Camp Forrest, whose barracks were erected around the monuments in honor of the Union and Confederate dead at the Battle of Chickamauga, was destined to train infantry engineers. In the same field, Camp Greenleaf was created in May 1917 to train medical officers for motorized or hypomobile units, evacuation and basic hospitals, as well as veterinarians and dentists. In 18 months, 6,640 officers and 31,138 soldiers passed through it. The three camps were abandoned in December 1918.

Responsible for preventing the spread of epidemics in the United States was

Other traces of the herald wave

Rupert Blue, born on May 30, 1867 in the county of Richmond, North Carolina, 60 miles east of the capital of that state, Charlotte, and on the border with South Carolina, where he would move shortly after with his parents to the hometown of his mother, Marion, another 60 miles south of his native county. His father was a Confederate Colonel and his older brother, Victor, naval officer in Annapolis and hero in the Spanish-American War, in which he fulfilled the task of monitoring enemy ships.

After being apprenticed in a pharmacy, Rupert Blue studied two years at the University of Virgina and graduated in Medicine from Maryland in 1892, then spent nine months as an intern at the Baltimore Naval Hospital, after which he was one of the four chosen, out of a total of 25 candidates, to enter the Public Health Service (USPHS) in March 1893.

Among his first destinations was Galveston (Texas), where he met the young actress Juliette Downs, who he married in 1895, although the nomadic life of the doctor - destined to the quarantine station of Angel Island in San Francisco - was cause that in 1902 they divorced. Overcoming this event at the time considered particularly unfortunate, the prestige of Blue grew after his success in the fight against yellow fever in New Orleans in 1905, his participation in the international exhibition of Jamestown (Virginia) in 1907 and especially his combat against the epidemics of bubonic plague in San Francisco in 1902-1904 and 1907-1908, supported by the recent discovery that rats spread the disease.

In addition to disinfecting 11,000 houses in the Californian metropolis and fill the concrete streets, and with the advantage that when he arrived at that post the epidemic had already been identified, Blue had to fight anyway, according to Mike Stobbe in his compilation of biographies of the heads of the USPHS (page 43), with its own character:

> He remained a shy boy at heart, and had to work
> at appearing comfortable in front of large audi-
> ences and speaking with enough skill and passion
> to stir people to adopt hygienic measures. (Even
> at his most rousing, Blue lacked the folksy talent

Santiago Mata: 1917 American Pandemic

possesed by Colby Rucker, his handsome and lively second-in-command in San Francisco, who became known as *Garbage Can Rucker* for his funny, popular presentation to women's groups about the proper disposal of trash.) But Blue was no slouch. Like Rucker, he gave half-a-dozen speeches daily and spent nearly as much time winning the cooperation of people as they did killing rats and other plague-spreading vermin. Together, the two gradualy convinced more and more people to clean up their properties and support their extermination campaign.

On November 21, 1911 died Walter Wyman, third US Surgeon General, and as such head of the USPHS. Rucker was responsible for presenting the candidacy of Blue to succeed Wyman in Washington. Blue was elected to the position by the 27th President of the United States, William Howard Taft (1857-1930, in office between 1909 and 1913), according to Stobbe (p.41):

President Taft was left to choose between two men. One vas Rupert Blue, the forty-six-year-old hero of the San Francisco plague outbreaks. The other was Joseph H. White, an older veteran of the Service (he was fifty-three at the time) who has distinguished himself on a variety of assignments, including the nation's last yellow fever outbreak in 1905. The selection process dragged out for more than a month, and Taft enjoyed building a little suspense –he jokingly invited reporters to pick their favorite color, Blue or White. The contest split opinion within the Service, with many officers considering White the more sensible choice. Among them was the brilliant epidemiologist Joseph Goldberg, who felt White had a moral courage that Blue lacked but believed Blue had the advantage of being more politically adept and –like Taft- a Republican.

Other traces of the herald wave

On the advice of Treasury Secretary Franklin MacVeagh, Taft in early January 1912 announced Blue won the job. The president also decided to limit the surgeon general to four-year appointments, instead of the unlimited terms Blue's predecessors had enjoyed.

Stobbe defines Blue (p.10) as "an amateur boxer with a twisted mustache, who agreed to the position because of the admiration he aroused" in combating the bubonic plague in San Francisco:

Blue was an instinctively quiet man who nevertheless built up the surgeon general's bully pulpit, leading health education campaigns and speaking out on the need for national health insurance.

To explain the admiration that Blue aroused, Stobbe (42) points out:

Blue was like a character out of a Rudyard Kipling novel: an imposing, mustached amateur boxer who had traveled the world in service of his country, reretaining his optimism through a series of sometimes lonely and discouraging assignments. A female reporter for the San Francisco Bulletin, profiling Blue in 1908, seemed to nearly swoon in midsentece: *A man of action rather than word – big, broad-shouldered, handsome, commanding in his plain brown officer's uniform. Yet modest and unassuming... a quaintly drawn Southern wit and kindly sympathy always lurking in his eyes and under the slim moustache, one to inspire confidence and to make one feel if worse should come to worse, we would have a man at the helm.*

Stobbe, however, concludes that "certain doctors gradually came to see Blue as a scientific lightweight, a public relations man who might do fine speaking to laymen but lacked the intellectual power to make timely decisions under challenging circumstances", and he mentions as "a particularly harsh assessment" the judgment made on him by the historian John Barry (p.242 of his

Santiago Mata: 1917 American Pandemic

book):

> Blue became surgeon general simply by carrying
> out assigned tasks well, proving himself an adept
> and diplomatic maneuverer, and seizing his main
> chance. That was all.

Properly, it was in August 1912, when President Taft gave the department of Blue its definitive name (U.S. Public Health Service), since until then it was called U.S. Public Health and Marine Hospital Service, extending its competences to "all human diseases" and not just some of the infectious diseases. Woodrow Wilson, new President since November 1912, was also a friend of Blue (according to Stobbe, "at least at first") and in June 1913 signed a decree by which the Service would receive an additional $ 200,000 for field work, an amount equivalent to the money needed to maintain the 33 naval hospitals of the Service and more than a tenth of the total budget of 1.7 million.

By then, the USPHS had a thousand employees, including 140 medical officers, controlling in addition to hospitals 50 quarantine hospitals and 125 relief homes, apart from a leprosarium in Hawaii and a tuberculosis sanatorium in New Mexico. In 1918 there were 180 doctors and 44 quarantine stations. Blue faced with particular effort the fight against pellagra, a disease derived from the low consumption of vitamin B3, first diagnosed in the United States in 1907 and which in 1915 affected between 25,000 and 100,000 people, so Blue commissioned Joseph Goldberger (1874-1929) to discover his causes, which this doctor of Hungarian origin born in what is now Slovakia made in December 1915, ruling that it was not an infectious disease but due to poor nutrition (for this reason, he would be nominated five times for the Nobel Prize in Medicine).

Also in 1915, Blue became the first US Surgeon General elected president of the American Medical Association (AMA), qualifying Blue as the doctor who had done the most for the good of humanity. In June 1916, according to Stobbe (p.46), Blue made at the annual AMA meeting in Detroit an apology for mandatory health insurance, which was already breaking through in the United Kingdom and Europe, as "the next great step in social legislation " to

Other traces of the herald wave

prevent deaths of poorly-fed workers. But trade unionists like Samuel Gompers (1850-1924) would oppose such expenses, preferring to claim wage increases.

In the face of the world war, Blue remained focused on domestic diseases, under the slogan *Safety First*, advertising in 1916 along with other state agencies the progress made with a mobile expedition composed of nine rail cars, visited daily by 6,500 people. As the United States' entry into the war approached, Blue's plan was to concentrate health care around the training camps of the army and navy, as well as in the factories of armaments and war material.

Faced with the 1918 flu, according to Stobbe (p.50), "Blue and his troops were slow to understand how large and severe the threat was. During the first wave, the Service did little to investigate reports in the late spring of flu death spikes in young people in Kentucky and a few other locations. " At the end (p.109) Blue would resort to spread "calming messages " that "proved absurd as it became clear that the Spanish flu was much deadlier than Blue had stated."

This last statement by Stobber, also anticipating the wrong name that the epidemic would receive, must be specified: When did those Blue messages about the flu begin? According to Christine M. Kreiser, Blue's first false step would have taken place in July 1918, when he rejected a direct request to allocate $ 10,000 to research on pneumonia (its purchasing value in 2018 according to *dollartimes.com* would be $ 180,000, equivalent to 154,000 euros).

This same author points out that Blue's first public warning about the flu would not arrive until mid-September, and would limit itself to advising under the slogan "try to make nature your ally" by not wearing dresses, shoes or tight gloves and that the food was selected and chewed well. Even when Congress assigned $ 1 million to USPHS to fight the flu, Blue would return $ 115,000.

Santiago Mata: 1917 American Pandemic

Beyond that, government agencies would resort directly to lying, Kreiser recalls:

> Worse still, the government contributed to the national paranoia surrounding all things German. The USPHS officer for northeastern Mississippi planted stories in the local papers that *the Hun resorts to unwanted murder of innocent noncombatants... He has [at]tempted to spread sickness and death thru germs, and has done so in authenticated cases.* Lieutenant Colonel Philip Doane, head of the Health and Sanitation Section of the Emergency Fleet Corporation, which oversaw U.S. shipyards, theorized that U-boats had delivered German spies to America to turn loose Spanish influenza germs in a theatre or some other place where large numbers of persons are assembled. So persistent was the belief that Germany had somehow launched a biological attack that USPHS laboratories devoted precious time to investigating claims that Bayer aspirin, which was manufactured in the States under a German-held patent, had been laced with deadly flu germs.

> *Let the curse be called the German plague*, declared *The New York Times* in October. *Let every child learn to associate what is accursed with the word German not in the spirit of hate but in the spirit of contempt born of the hateful truth which Germany has proved herself to be.*

For Barry (p.135), part of the failure could be due to the blockade that the physicians who controlled the medical committees of the National Research Council and the Council of National Defense, namely William Welch, William Gorgas, Vaughan and the brothers Charles and William Mayo (the five had been presidents of the AMA), exercised against Blue for considering him incompetent:

Other traces of the herald wave

Welch and his colleagues so doubted his habilities and judgment that they not only blocked him from serving on the commitees but would not allow him even to name his own representative to them. Instead they picked a USPHS scientist they trusted. It was not a good sign that the head of the public health service was so little regarded.

From the beginning of their planning, these men focused on the biggest killer in war –not combat, but epidemic disease.

William Henry Welch (1850-1934), recognized as a key figure in American medicine, the son of a Connecticut rural doctor and motherless since he was six years old, enrolled at the Winchester Military School aged 13 and aged 16 at Yale University, where he graduated in 1870. In 1875 he obtained the title of doctor of medicine, and completed his training in Strasbourg (then in Germany), where he studied histology with Heinrich Wilhelm Gottfried Waldeyer (1836-1921), chemistry with Felix Hoppe-Seyler (1825-1895) and pathology with Friedrich Daniel von Recklinghausen (1833-1910), discoverer of neurofibromatosis (1882) and hemochromatosis (1889). He expanded his physiology studies in Leipzig with Carl Friedrich Wilhelm Ludwig (1816-1895) and Hugo Kronecker (1839-1914), specializing in the study of pulmonary edema in Breslau with Julius Friedrich Cohnheim (1839-1884).

Already an expert and passionate about bacteriology, he inaugurated the first laboratory of physiology and pathology in the United States at the Bellevue Hospital in New York. John Shaw Billings (1838-1913), who had known him in Germany, proposed him as one of the founders of a new hospital in Baltimore, so Welch returned to Europe to catch up on bacteriology with the doctors Karl Georg Friedrich Wilhelm Flügge (1847-1923), Heinrich Hermann Robert Koch (1843-1910) and Max Joseph von Pettenkofer (1818-1901) and on his return he joined the laboratory of British biology professor Henry Newell Martin (1848-1906).

Since 1889, Welch worked at the Johns Hopkins University Hospital in Baltimore, and from 1893 to 1898 he was the first dean of the School of Medicine,

Santiago Mata: 1917 American Pandemic

discovering in 1891 the bacillus that will bear his name (or that of *Clostridium perfingens*). He left the chair in 1916 to found the School of Public Health (School of Hygiene and Public Health at Johns Hopkins) with funds from the Rockefeller Foundation and this in 1925 to found a department of History of Medicine and the Library that bears his name. In 1896 he founded the *Journal for Experimental Medicine* and in 1920 the *American Journal of Hygiene*.

When he joined the army during the First World War, Welch was already 67 years old and had just left the position of president of the Academy of Sciences (1913-1916). In his 13 months as brigadier general, his task was to review the hygienic conditions of the troops that were to be sent overseas. Supported by his relationship with the 22nd Surgeon General of the U.S. Army (1914–1918) William Gorgas, he incorporated doctors such as the Mayo brothers, George Crile and Harvey Cushing into the service.

William Crawford Gorgas, born in Alabama in 1854 and died in 1920, become famous as responsible for fighting yellow fever and malaria in Florida, Cuba (1898) and the Panama Canal (1904), proving that both were transmitted by mosquitoes.

Also Southern but from Missouri was Victor Clarence Vaughan (1851-1929), doctor of Medicine by Michigan in 1878, where he was dean since 1891. Militarized for the Cuba campaign, he became familiar there with diphtheria, which earned him the title of father of bacteriology in the United States.

As for the brothers Will and Charlie Mayo, they were the main exponents of the Mayo Clinic, founded in Rochester, Minnesota, in 1889 by nine people: William Worrall Mayo, William James Mayo, Charles Horace Mayo, Augustus Stinchfield, Christopher Graham, Henry Stanley Plummer, Melvin Millet, E. Star Judd and Donald Balfour. The first (1819 - 1911), descendant of the English chemist John Mayow (1641-1679) was the main promoter but by his early death excluded from the seven founders (the former ones except him and Graham), who in 1919 created the Mayo Properties Association.

His children William James (1861 - 1939) and Charles Horace (1865 - 1939) graduated in medicine before working in his father's office and then in the

Other traces of the herald wave

hospital. William was a colonel in the department of the Surgeon General and his brother Charles was in the Army medical corps, also acting as William's substitute in the position of associate chief delegate for the surgical care of Army soldiers.

In 1916, President Wilson organized the Committee of American Physicians for Medical Preparedness, chaired by William and with his brother Charles as a member. This committee was transformed into the General Medical Board of the Council for National Defense, in whose executive committee William was a member, with Charles as his deputy.

The Mayo Clinic participated in the training of the military doctors, always staying in the Clinic one of the two brothers - who would end the war with the position of brigadier general and the distinguished service medal - while the other was in Washington.

Given the importance of this epidemic, the question arises as to whether doctors and US authorities could identify it with the same speed with which its seriousness extended, and if so, whether they reacted in the manner in which it could be expected given their scientific knowledge and positions they occupied. First it is necessary to know the disease and study to what extent its causes, symptoms and effects were known by doctors and health authorities, so that they could take appropriate measures.

We have already anticipated by citing Hoffman's study that the pandemic flu of 1918 was caused by the H1N1 virus and that it caused in healthy adults, which would be its main group of fatalities "a storm of cytokines", while older individuals enjoyed some immunity for having been previously exposed to a similar virus (the flu of 1889) and the youngest for a strange phenomenon of immunological "honeymoon".

The virus of the Russian or Asian pandemic of 1889 was not H1N1 but H2N2, although it is doubtful whether it could be H3N8. The pandemic reached the United States in 70 days and in four months it had gone around the world, with a mortality rate of 1% of the patients, resulting in approximately one million fatalities.

Santiago Mata: 1917 American Pandemic

The influenza viruses comprise three genera: A (of avian origin), B (which affects only humans) and C (which infects humans and pigs). Together with two other genera (*Isavirus*, cause of infectious salmon anemia, and *Thogotovirus*, transmitted by ticks) are of the *Orthomyxoviridae* family, which in turn is part of the group of ribonucleic acid (RNA) V viruses, called single-stranded negative (virus (-) ssRNA) by having a single chain and being the sense or polarity of its negative RNA, which means that it is necessarily infectious because it needs to be translated (converted) into positive RNA by an RNA polymerase.

RNA polymerases are proteins (molecules formed by amino acids, which in turn are organic molecules with an amino group -NH2- and a carboxyl group -COOH) enzymatic (accelerating chemical reactions) capable of forming RNA from double chains of DNA, dividing them into thousands or millions of nucleotides (organic molecules of pentose-five-carbon monosaccharide-plus a nitrogenous base and a phosphate group). The main polymerase is what allows the synthesis of messenger RNA (mRNA), which contains the genetic information sent from the nucleus of a cell to a ribosome in the cytoplasm to determine the order in which the amino acids in a protein will bind.

The viruses of the *Orthomyxoviridae* family are the only ones of the V group that carry out their replication inside the cell they attack (and not in the cytoplasm, like the rest). This multiplication process has six stages:

1. Transcription of the negative RNA by the RNA polymerase within the virion (complete and infectious virus particle) and production of mRNA and positive single-stranded RNA.

2. Translation of mRNA and production and accumulation of early (regulatory) proteins.

3. The regulatory proteins encourage the production of positive RNA and negative RNA.

4. Late transcription of the negative RNA.

5. Delayed translation of the negative RNA and production and accumulation of late (structural) proteins.

Other traces of the herald wave

6. Assembly of the nucleocapsid (protein cover enclosing the nucleic acid) and maturation.

While this process is common to all viruses of group V, *Orthomyxoviridae* are differentiated because they have wrapping and rod or elongated form (they are isometric). From the inside to the outside, they have five components:

1. Ribonucleoproteins: (RNP) are nucleoproteins (NP, proteins associated with a nucleic acid, which can be DNA but in this case is RNA), combining ribonucleic acid with proteins in 8 segments of single-stranded RNA in the types Influenzavirus A and Influenza virus B (C has 7 segments). RNP is a specific antigen (substance that triggers an immune reaction), which distinguishes between types A, B and C.

2. Polymerase (RNA-Pol) or reverse transcriptase. Enzymes to synthesize viral RNA. Segments 1, 2 and 3 of the RNA of Influenza A, B and C viruses encode three Pol-RNA proteins.

3. Non-structural proteins (NS1 and NS2): segment 8 of viral RNA encodes at least two non-structural polypeptides (molecules with more than ten amino acids) that are translated by different mRNA. The human, porcine and equine NS1 are indistinguishable from each other and those of avian viruses give rise to multiple immunological responses (they present antigenic variability). The NS2 protein is synthesized at a late stage.

4. Proteins M (Matrix or Membrane): segment 7 of the virus encodes the M1 and M2 proteins. M1 is a structural protein of the virion, associated with the double fat layer (lipids) of the virus envelope, so it is very close to the surface glycoproteins and the RNP complex. Like RNP, it is a specific antigen of the virus type. M2 is considered a non-structural protein.

5. Membrane glycoproteins (cover proteins or viral envelope): Hemagglutinin (HA) and Neuraminidase (NA). HA represents 25% of total virion proteins and NA 6.7%.

Santiago Mata: 1917 American Pandemic

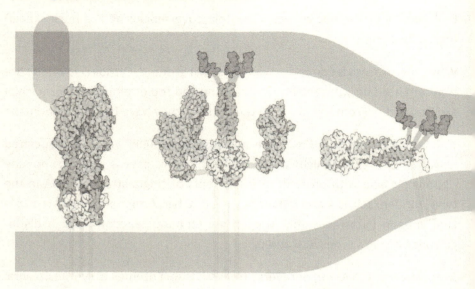

(Process by which Hemagglutinin binds the walls of the cell - above - and the virus - below -, graphic by David Goodsell, http://pdb101.rcsb.org/motm/76)

Hemagglutinin is projected from the lipid membrane of the viral envelope, like a needle (spicule) of approximately 14 nm (nanometers, billionths of a meter) by 4 nm. It is evenly distributed on the surface of the virion, whereas NA can be concentrated in certain areas. The functions of HA are to participate in the adsorption (anchoring) and penetration of the virus into the cell, stimulate the fusion between the membrane of the host cell and the viral envelope, agglutinate the red blood cells (erythrocytes or red blood cells) producing a reaction of visible agglutination and induce the synthesis of neutralizing antibodies.

Neuraminidase: is a glycoprotein or receptor-destroying enzyme, which extends from the viral membrane as a projection or spicule formed by a quadrangular head. Its functions are to provoke (catalyze) the division (cleavage) of the bonds between the sialic acid (N-acetylneuraminic, monosaccharide derived from neuraminic acid) terminal (whose carbon 2 can bind to carbon 3 or carbon 6 of galactose, sugar that plays an important role regarding the transmission of influenza between species and its mutation) and an adjacent

Other traces of the herald wave

sugary cell residue that allows transporting the virus through the mucins (high molecular weight proteins and fats, which make up the mucous membranes) and destroy receptors of the HA on the host cell, allowing the extraction (elution) of the viral progeny of the infected cell, preventing viral aggregation (fusion), protecting the virus from its own HA, stimulating the production of neuraminidase inhibitory antibodies (molecules that bind to the enzyme and decrease its activity).

According to Raquel Almansa, while the genera of influenza B and C have a species each without subtypes, 18 subtypes of the species of influenza A (unique to the genus Influenza A) have been described, based on variations of Hemagglutinin and Neuraminidase : H1N1, H1N2, H2N2, H3N1, H3N2, H3N8, H5N1, H5N2, H5N3, H5N8, H5N9, H7N1, H7N2, H7N3, H7N4, H7N7, H9N2, H10N7.

This same author explains the mutations of influenza viruses through "a process of continuous evolution mediated by mutations that allows them to adapt to the host and environmental conditions. It is an obligatory evolution known as selective pressure ", whose mechanism consists of "nucleotide substitutions, deletions and insertions of triplets caused mainly by the errors of the RNA polymerase during transcription. These mutations accumulate in each cycle of transcription and generate variants of protein function or structure. "

The mutation frequency of influenza viruses is one per hundred thousand nucleotides and the new variants benefit from the fact that the antibodies already created can not neutralize the new virus. Minor variations (called antigenic drift) give rise to the different annual flu epidemics. Larger variations, in which a true antigenic replacement occurs, if they include changes "in genes that code for HA or NA" could "spread rapidly and cause great morbidity and mortality in the entire population, including young and healthy people, since the previous existing immunity would not be able to stop the development of the infection. " For the origin of the major variations, Almansa cites two hypotheses:

Santiago Mata: 1917 American Pandemic

Genetic rearrangement: This process consists of an exchange of genes from viruses from different species that coinfect the same host. For the rearrangement to be carried out, there must be a species susceptible to being infected by both viruses. Human influenza viruses have preference for sialic acids bound to galactose by an alpha-2-6 bond. It is believed that this is the biochemical basis by which viruses can not cross inter-species barriers, since avian viruses have preference for sialic acids bound to galactose by an alpha 2-3 bond.

In the pharynx of pigs there are both types of bonds between sialic acid and galactose, which is why they are susceptible to infection by avian and human lineages, favoring the appearance of new viral strains potentially pandemic by rearrangement between avian and human viruses. This would explain the fact that pandemics often arise in geographical areas where humans, pigs and waterfowl coexist in close proximity.

The second hypothesis would be the "direct jump of species barrier: It is a direct transmission to humans without intermediaries. Most of these inter-species transmissions do not thrive and simply cause occasional isolated outbreaks. "

In the case of the influenza of 1917-1918, the second hypothesis seems more likely, since its radical novelty surprised the immune system of young humans (the possibility of a certain immunization remaining for those who had been influenced by the flu of 1889). Given that after the flu went from humans to pigs, it is possible that in the mutation between the first and second waves of the pandemic occurred the genetic rearrangement pointed out in the first hypothesis.

Before looking at this issue more extensively, let's look at the particularities of the replication of the influenza virus, as Almansa relates:

Other traces of the herald wave

The influenza virus mainly infects epithelial cells of the upper and lower respiratory tract. The viral particles adhere to their cellular receptor, the sialic acid, which is recognized by the HA of the envelope or envelope membrane. The binding to the receptor initiates the absorption of the viruses through the endocytosis [entrance of molecules in the cell] mediated by receptors. A vesicle is thus formed [an ampoule or pouch] that fuses with the intracellular compartments called endosomes [membrane vesicles].

Thanks to the low pH inside the endosomes, maintained by means of the proton pumps inside the membrane, the fusion reaction between the viral envelope and the endosomal membrane is triggered. The low pH of the vesicles reaches the interior of the virion through the ion channel formed by the M2 protein, which generates a destabilization of the interaction of the RNPs [ribonucleoproteins] with the M1 protein and a change in formation of haemagglutinin, which favors the fusion of the envelope of the virus with the endosomal membrane and the formation of a fusion pore through which the genetic material enters the interior of the cell. The RNPs are released into the cytoplasm where the processes of transcription and replication of the viral genome take place.

The RNP complexes are transported to the nucleus carrying with them the transcriptase [polymerase] formed by PB1, PB2 and PA. Transcriptase by a process called *head theft*, cleaves small regions of the cap of the cellular mRNA and uses them as baits for the synthesis of viral mRNA. This viral mRNA is transported to the cytosol [liquid inside the cell] where the protein translation will take place.

Santiago Mata: 1917 American Pandemic

The virus uses all the machinery of the cell for the synthesis and processing of its proteins. HA and NA, through the Golgi apparatus [organelle of the cell that serves to make proteins], reach the plasma membrane [the last or outer cell]. The NPs [nucleoproteins] and the RNA polymerase components interact with the viral RNA synthesized to form the RNPs that interact with the M1 that covers them. The RNPs travel to the cellular plasma membrane through microfilaments of actin [globular protein] and arrive at regions rich in cholesterol called *lipid raft* in which the NA and HA are anchored. The viral particles leave the cell by gema- tion or budding [development of a bud], acquiring the lipid envelope in which the glycoproteins were anchored. Finally, the NA hydrolyzes [forms an acid and a base from a salt by interaction with water] the receptors that the virus has dragged from the cell membrane, which prevents them from being added and allows their dispersion by the respiratory tract.

In their 2007 article on the discovery of the H1N1 virus of the 1918 pandemic, Taubenberger, Hultin and Morens state that "at the time of the 1918 influenza pandemic, no-one suspected that the cause of the human disease was derived from an avian infectious agent. " The *archeological* investigations of 1951 and 1995 allowed to recompose the eight genes of the pandemic virus, which were very similar to the avian but also different from the viruses of that origin circulating at that time, which otherwise would not have changed much in the next ninety years. For those three authors, the origin of the virus is mysterious:

Viral sequence data now suggest that the entire 1918 virus was novel to humans in, or shortly before, 1918, and that it was not likely to have been a reassortant virus such as those that caused the 1957 and 1968 pandemics. Rather, the 1918 virus is an avian-influenza-like virus that appears to

Other traces of the herald wave

have been derived *in toto* from an unknown source because its eight genome segments differ from contemporary avian influenza genes, especially at synonymous sites. Influenza virus gene sequences from a number of fixed specimens of wild birds collected circa 1918 showed little difference from avian viruses isolated today and consequently did not suggest these birds were the source. These findings also suggest that avian viruses undergo little directed evolution in their natural hosts even over long periods.

Certainly, ignorance about the virus did not imply lack of means to prevent its expansion or to fight its symptoms. And even though the virus was as deadly as H1N1, when infecting monkeys, it multiplies the mortality rate in relation to any other type of human influenza virus.

Apart from the doubts expressed by Taubenberger and his colleagues in 2007, others concluded that there was a direct transfer of H1N1 from bird to human. This is the case of Robert Belshe in the work he published in 2005:

> The startling observation of Taubenberger et al. was that the 1918 virus did not originate through a reassortment event involving a human influenza virus: all eight genes of the H1N1 virus are more closely related to avian influenza viruses than to influenza from any other species, indicating that an avian virus must have infected humans and adapted to them in order to spread from person to person. Thus, pandemic influenza may originate through at least two mechanisms: reassortment between an animal influenza virus and a human influenza virus that yields a new virus, and direct spread and adaptation of a virus from animals to humans.

Santiago Mata: 1917 American Pandemic

The "first" wave of flu in North America

I consider proven that the first wave of pandemic flu emerged within the ordinary time frame of the autumn flu, that is, from October 1917. Because of the novelty of this discovery, I shall call the initial outbreak a *herald wave*, according to the proposal of Hoffman, respecting that it is considered as the first wave the one which manifested itself virulently in March 1918.

The extension of that usually called first wave of the 1918 flu in the United States coincided with the change in the leadership of the Army General Staff and the challenge of sending troops to Europe, that is, with the dilemma of whether or not to send the troops knowing they were sick.

Peyton C. March (1864-1955) was elected chief of staff on March 4, 1918, the same day when the epidemic broke out in Camp Funston, Kansas. The date was fictitious, because March's official appointment did not arrive until May 20 . But in March he assumed the position temporarily, being promoted to acting general.

Born in Easton (Pennsylvania), Peyton March graduated as an officer in 1888. Married since 1891 to Josephine M. Cunningham, he was stationed at the Fort Monroe Artillery Academy until 1898, in the Philippines he commanded the American forces on December 2, 1899 in the battle called of the clouds, in the Paso de Tilad, Luzon. He was promoted to lieutenant colonel, was civil and military governor of several Philippine provinces and general commissar of prisoners, assisting as a military observer to the Russo-Japanese war of 1904, already at the service of the newly created staff of Washington.

In 1917 March was appointed chief of the artillery of the American Expeditionary Forces (AEF), whose commander, four-star general John J. Pershing (who preceded him in rank) would oppose March's appointment as chief of staff.

The work of March is considered by scholars as essential to establish the existence of the staff and its head as the supreme command of the US Army. When he was appointed, however, the main challenge was to respond to Pershing's requests to send more troops. There is no evidence that any doctor asked that measures be taken to prevent the spread of the influenza.

Santiago Mata: 1917 American Pandemic

The official starting date for the epidemic in Camp Funston -where 56,000 recruits were crammed- is Monday, March 4, 1918, when the cook Albert Gitchell was ill with a fever of 40 degrees, which according to Honigsbaum (p. 41) makes it for "many scientists" (at least those who ignore what happened in 1917) the zero patient of the pandemic flu. The next was Corporal Lee W. Drake, truck driver from the transport headquarters of the first battalion. Sergeant Adolph Hurby followed, with 41 degrees of fever, nasal and bronchial inflammations.

With those three cases, the doctor of the infirmary called the health chief colonel of the camp, Edward R. Schreiner, 45 years old. That night, his hospital with 3,068 beds already had 107 hospitalized by the epidemic, which at the end of the week (10 was Sunday) were 522. Schreiner, who found the *Pffeifer bacillus* in several cultures taken from the sick, waited another three weeks before communicating by telegram to Washington on March 30, the same day that according to the number of Public Health Reports published on April 5 the Haskell flu had been reported (as quoted by Honigsbaum, p. 43):

> Many deaths influenza following immediately two
> extremely severe dust storms.

To keep track of the flu among civilians, we must draw on scarce information. On the other hand, there is more information about its spread in the network of military training camps that trained and sent soldiers to New York and, from there, to Europe.

In Camp MacArthur, the Volume XII of *The Medical Department* indicates (p.108) for April 1918 a number of hospital admissions for acute respiratory diseases only slightly higher than in January (more than 700), with only a dozen cases of measles and one death. In May there were two deaths, with an increase in measles (almost 40 cases) and a decrease in admissions below 600; these were in June only 150, with something more than 20 of measles and again two deaths.

In Camp Cody, where we saw an epidemic of influenza in the fall of 1917 (and

The 'first' wave of flu in North America

a mortality rate of cases of serious respiratory diseases increasing between November and December from 0.95% to 1.71%), the Volume XII of *The Medical Department* records (p.31) that "influenza cases increased in March, 1918" (therefore, the epidemic had not ceased), "without increase in case fatality", and that a "sharp rise in June followed the receipt of between four and five thousand men, and the streptococcus again was prevalent at this time, the case fatality rate being 1.95% for this month, " then dropping in July to 0.46% and increasing in August to 3.73%, "preceding any apparent clinical evidence of the beginning of an epidemic," as if the mortality rate was not enough evidence, or perhaps showing that it was an epidemic of influenza without the streptococcus that doctors expected to camouflage it as pneumonia.

Compared to the *herald wave* of pandemic influenza in December 1917 at Camp Cody, it is not correct to speak of an increase in influenza from February to March, as admissions for respiratory diseases decreased (from 350 to 280) and the increase in April (300) was minimal. The rebound in June (in May there were only 150 admissions and three deaths) resulted in more than 250 admissions and five deaths: which also does not mean an increase compared to the period from November to April.

The nearest camp to Camp Funston was Camp Dodge, which was as far from Camp Funston as Haskell County, but in the opposite direction: it was less than 240 miles in a straight line, but to get from one camp to another by rail probably meant taking the train, rather than through Kansas City, from where there was no line to Des Moines, from Camp Funston (Riley on maps) to Belleville (on the line to Denver) and from there through Lincoln and Omaha to Des Moines (about 340 miles).

There was no information about this first wave at Camp Dodge until a small note appeared in the weekly *The Tomahawk* (published in White Earth, Becker County, Minnesota, 375 miles north of the camp), on June 27, 1918 (in the page 2 and dated on day 21), where it is indicated that the division present there (88th) had lost its normal state of health in March:

Santiago Mata: 1917 American Pandemic

> In less than two months the 88th national army division here, where Minnesota soldiers are in training, has regained its standard as one of the healthiest in the national army. Just prior to April 1, a pneumonia epidemic struck the division, causing the death of nearly 100 soldiers. This scourage was curbed early in May with the result that only 55 deaths from all causes were reported for the month as against 112 for April.

As we have seen, Camp Dodge had already registered more than a thousand hospital admissions in March, with more than 30 deaths, which would reach 40 in April (*The Tomahawk*, as we saw, doubles this figure to 100), although admissions declined below 1,000. Still in May there would be almost 500 admissions and 8 deaths (according to *The Tomahawk* they were about 50). In June there were less than 250 hospital admissions, without deaths. "The case fatality for all respiratory diseases which had been below 1% rose to 2.63% in the month of March and increased sharply during the latter part of this epidemic wave, the rate for all respiratory diseases for the month of April being 3.84% ", an increase due to the" increased incidence of streptococcus " associated with a "change in the type of pneumonia. "

After reducing the importance while calling it "uncomplicated influenza" , the Volume XII of *The Medical Department* (page 71-72) finally gives us the data on influenza cases among hospital admissions at Camp Dodge: 939 in March (77% of the total of 1,213 admissions), 136 in July and 28 in August.

Despite its proportions - a hundred deaths in front of the 17 American soldiers recently died in military action in France, according to the military report appeared on the same day and page in *The Tomahawk*, the epidemic of influenza in Camp Dodge was disguises as pneumonia and could not interfere in the propaganda work then underway: in that same page the newspaper reports that to promote the sale of war bonds, probably President Wilson himself was going to replace the Secretary of the Treasury in the trips to make propaganda in the presence of gathered multitudes, something incompatible with an epidemic that should be fought with quarantine measures.

The 'first' wave of flu in North America

In the case of Camp Pike, the epidemic that broke out in October 1917 and worsened in November and December, went from the minimum of 500 admissions in February to 750 in March, with few cases of measles (35) and 15 deaths. In April, it fell below 600 admissions, with measles in 40 cases and eight deaths. In May, for the first time since its inception, the epidemic fell below 400 admissions (nine deaths), a trend that was consolidated in June, when the admissions dropped under 300 (five dead), and went up slightly in the following two months.

In Camp Bowie (Texas), the Volume XII of *The Medical Department* recognizes (it must be remembered that it was written in 1929) that what the documents describe was an epidemic of influenza which the responsible doctors did not want to call as appropriate and that the recruits were infected as soon as they arrived (pages 24 and 26):

> Between March 26 and April 13, 349 cases of an acute respiratory infection were admitted. These were not called influenza because of lack of bacteriologic support for such diagnosis. Clinically, they closely resembled the condition subsequently called influenza, although this diagnosis is questionable.

> A definite influenza outbreak occurred in April and, though the streptococcus appears to have been prevalent, the case fatality was only 0.45% of the total respiratory disease. Recruits arriving at the camp picked up the prevailing infection with the streptococcus and other organisms. The bacteriological examinations in June showed the flora present in the interepidemic period and preceding the wave of the pandemic.

East of Lake Michigan, in Camp Custer, after falling in February 1918, mortality among those suffering from acute respiratory diseases increased in March to 1.67% (about 650 cases and 11 deaths) and in April up to 2% (cases fell below 600, but with 12 deaths). The Volume XII of *The Medical Department*

Santiago Mata: 1917 American Pandemic

recognizes again the manipulation of the records (p. 39):

> This wave of respiratory diseases was definitely influenzal in nature, thought the deaths were attributed to or diagnosed as primary pneumonia. Here, as in other camps, while the cases may have been admitted with influenza or acute respiratory disease, the diagnosis was frequently changed to pneumonia and the deaths were so recorded.

On the other side of the lake, in Camp Grant, after the already mentioned epidemic with peak morbidity in January (700 cases) and peak mortality in February, in March there were three deaths, with mortality rebound in April (16), although morbidity was not striking (400 admissions).

In Camp Sherman (Ohio), as we saw when talking about the beginning of the flu in 1917, there was a continuity between the *herald wave* and the one called first, going from 800 admissions and four deaths in February to more than 900 in March (only 30 cases of measles) and more than a dozen deaths. In April admissions fell, but remained above 800 (only 15 cases of measles) and deaths were eight, remitting the epidemic noticeably in May (less than 400 admissions -25 cases of measles- and one death) and June (100 admissions, four for measles, and no deaths).

Also in Camp Taylor, after a decline in February, the epidemic rebounds in March with 750 cases (more than 100 are measles) and more than 15 deaths, rising in April to 1,000 hospital admissions (100 measles) and 12 deaths, falling in May to something more than 400 admissions (50 measles) and five deaths, and arriving in June 1918 to its lowest point with 150 admissions (again almost 50 measles) and no deaths.

On Camp Devens, the figures of the first wave were published in the Volume XII of *The Medical Department* (page 42-44):

> There was a sharp rise in the case fatality rate in March, the rate being 1.07%, and increasing to 1.97% in the month of April. This increase in deaths is ascribed to primary pneumonia. The

The 'first' wave of flu in North America

character of the cases, as shown in the histories and in the protocols of necropsies, makes it quite probably that an influenzal wave occurred at this time.

In this camp, as in several others, a rise in the total fatality rate of all respiratory diseases preceded the wave of influenza. The case fatality rate was 0.61% in June, 1918, rising to 1.48% in July and to 3.83% in August, while the definite wave of the pandemic in not recorded as having started until September.

With an average force of 29,613 soldiers between September 1917 and May 1918, there were more than 600 patients in Camp Devens and 7 died of influenza or other respiratory diseases in March 1918, 500 patients and 9 dead in April, less than 200 patients and 3 dead in May, 150 sick and two dead in July, 150 sick and five dead in August, so it could almost be said that in Camp Devens the flu remained in its severity from March to August, with peaks in March and April.

Perhaps the only military camp that was certainly spared the first pandemic phase at the end of 1917 was Camp Fremont, in Menlo Park, near Palo Alto (California), south of San Francisco Bay (almost 890 miles northwest of Camp Cody and 666 south of Camp Lewis), for the simple reason that it was opened in January 1918. The Volume XII of *The Medical Department* claims (p.75) that "respiratory diseases in epidemic proportions did not occur until late ":

The epidemic rise, with peak in April, was due to influenza and common respiratory diseases, the case fatality being low. Case fatality for all respiratory diseases increased prior to the onset of the influenza epidemic.

Although this reference avoids linking the epidemic of April with the second pandemic wave that will explode in Camp Fremont on October 8, the data provided -350 admissions in January 1918, 45 of them for measles, without deaths- allow us to know that in April there were more than 1,000 admissions

69

Santiago Mata: 1917 American Pandemic

(in March there had also been 350), only 35 for measles, and four deaths. In May, admissions fell back below 400, with one death.

After having referred to the second epidemic wave in this camp, the same source will refer to the first wave (p.81) in contradictory terms with the just mentioned:

> The influenza epidemic of the spring of 1918 was accompained by considerable pneumonia, with a sharp rise in the mortality.

Camp Lewis, who had registered in November and December of 1917 a *herald wave* with remarkable morbidity (550 admissions each month) but little relevant mortality, sees the flu intensified in March 1918 with 900 hospital admissions (60 cases of measles) and only one or two deaths per month until June (meanwhile the new admissions fall to 800 in April and 300 in May, to end with an atypical month in June whit only 250 admissions (20 measles) but four deaths.

At the end of June, between the 21st and the 24th, after nine months of training, the 27,000 men of the 91st division left Camp Lewis for Camp Merritt (in Cresskill, Bergen County, New Jersey), 14 miles north of the Port of Hoboken, where they embarked as of July 6 to England, except for some units that went directly to France.

About Camp Merritt says the Volume XII of *The Medical Department* (p.114) that, being a boarding and disembarkation camp through which troops from all over the United States were passing, "records from which a graph could be made are not available".

As for how to fight against the flu, then like today there is still only one effective way: isolation, avoiding all contact with the sick. Although it was not known then that the flu was transmitted through the air because of sneezing and mucus, the quarantine rule was known, and examples were given (as we will see in Alaska) of populations that were saved from the pandemic by that method, while neighboring localities were decimated.

The 'first' wave of flu in North America

It is not possible to determine if the responsible doctor of Haskell County, Loring Miner, raised immediately the alarm to the higher authorities, or if he did it on March 30 as stated on April 5 in the *Public Health Report*. There were reported 18 cases of "severe influenza" which resulted in three deaths, one for every six patients or a mortality of 16.7%, incredible if there had been really only 18 people infected.

In any case, Miner was not able to prevent the spread of influenza, nor is it likely that he was nothing more than the spectator of a process (contagion from Camp Funston to Haskell) inverse to the one that has been presented to us (contagion from Haskell to Camp Funston and from there to the rest of the world).

Officially, the communication of the epidemic was registered a month and a half (March 30) after it was reported in the *Santa Fe Monitor* (February 14). Between that date and April 5, when the *Public Health Report* announced the outbreak, phenomena had occurred that guaranteed the spread of the epidemic in the United States. The data of the *herald wave* suggest that these phenomena poured oil, or even gasoline, on a bonfire that had existed for months.

The first, already known, is the recruitment and training of soldiers, and their shipment to Europe. The second had also to do with the war, and would facilitate the spread of the virus among civilians: the great rallies to raise funds to pay for the war effort, through the so-called Liberty Bonds or Liberty Loans, which would make possible the loans of the United States to the Allied from the moment in which this country entered the war against Germany and the other Central Powers (April 6, 1917), and which were issued in four series, two in 1917 (on April 24, authorizing the issuance of 1,900 million of dollars in bonds at 3.5%, and on October 1, with 3,800 million at 3%) and two in 1918: on April 5 (4,100 million at 4,15%) and September 28 (6,900 million at 4.25%).

The British interpretation of the origin of the epidemic was published in 1920 in a report by George Newman (Chief of the Medical Cabinet of the Ministry of Health of the United Kingdom) to the Minister Christopher Addison. This

Santiago Mata: 1917 American Pandemic

Report on the pandemic of influenza 1918–19 states on page 283 that:

> Influenza appeared in the American army during February and March 1918. One of the first reports on the subject dealt with an outbreak about mid-March at Fort Oglethorpe, which appears to have coincided with the outbreak of influenza in the American Expeditionary Force, as well as in the British army in France, and in the French civil and military population. The outbreak at Fort Oglethorpe lasted three weeks, and 1,468 cases were admitted to hospital with influenza, out of a total strength of 28,586 men. There were many cases of so mild a type that they were not sent to hospital. The average incubation period was three days. Inquiry showed that previous to the outbreak there had been sporadic cases occurring for some time, and that the disease had been present in 1917.

Paradoxically, while the flu spread through the training camps, in Camp Uton (Long Island, New York), the vaccine against pneumonia created by Rufus Cole, director of the Rockefeller University Hospital, was tested on 12,000 recruits.

American Colonel Leonard P. Ayres stated in the war statistics whose second edition he published in 1919 (Chapter IX: *Health and casualties*, p. 125) that 40,000 soldiers died of "pneumonia" and "of these, probably 25,000 resulted from the influenza-pneumonia pandemic, " but that before September 14, 1918, "only 9,840 deaths from disease had occurred in the Army, and the death rate for the period of the war up to that time was only 5 per year for each 1,000 men. "

Nor is much more specific the document on infectious diseases published in 1928, prepared under the direction of Major General Merritte Weber Ireland (1867-1952, the 23rd U.S. Army Surgeon General from October 4, 1918 to

The 'first' wave of flu in North America

May 31, 1931) and drafted by Lieutenant Colonel Joseph F. Siler (military doctor) as Volume IX of *The Medical Department of the United States Army in the World War*. The 17 physical volumes of the 15 parts of this work have as editor in chief Charles Lynch and under that name appear in the bibliography.

The second chapter, dedicated to respiratory diseases, was written by Major Milton W. Hall. It occupies pages 61 to 171, and passes directly from the seasonal flu of 1917 to the "fall outbreak of 1918", thus ignoring both the *herald wave* and the first wave of pandemic flu. However, the data provided in tables 15 to 22 leave no doubt about the existence of the first wave of pandemic flu among US troops in the United States in March and April:

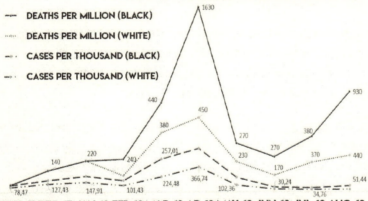

The same can be said for the American troops sent to Europe (American Expeditionary Force, AEF) between April and June:

THE FIRST WAVE OF PANDEMIC FLU AMONG US. ARMY SOLDIERS IN EUROPE (AEF)

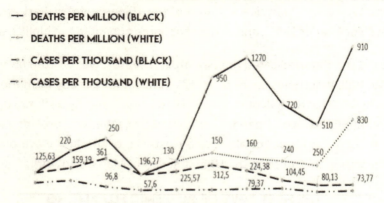

What happened with the US soldiers shows the transfer of pandemic flu from its place of origin, in the United States, to Europe. As we shall see, however, the United Kingdom will pretend to assume that the disease was imported there by its own soldiers repatriated from France.

But before Europeans, the first country infected by pandemic flu seems to have been Canada, where according to Osborne "an usually severe respiratory illness was circulating in the winter and spring of 1918" (p. 74). At St Luke Hospital in Ottawa there were 50 admissions of patients with flu symptoms (one of whom died) in January and February of 1918, a number that fell in March to reappear in April with a peak of 49 admissions (and two deaths by pneumonia), which is unusual, because between January and April 1917 there were only 28 admissions with flu-like symptoms (and no deaths).

The military hospital in the barracks of St-John (Saint-Jean-sur-Richelieu, in front of Montreal), also registered a first peak of 23 admissions for influenza, bronchitis or pneumonia in the first three weeks of January 1918 and another in April "that did not result in as many admissions but that was sustained over a longer period of time. " The admissions from flu-like illnesses from December 5, 1917 to March 4, 1918 were 63 (0.7 a day), from March 5 to June 4 were 92 (1.02 a day). In summer (June 5 to September 4), despite of

The 'first' wave of flu in North America

ships with cases of influenza in the port of Quebec, they dropped to 36 (0.4 per day).

A report by J.A. Amyot, health consultant to the director general of medical services of the Canadian army, identified in April the flu in that country's soldiers stationed in England. Reports from the front, according to Osborne, indicate that it was already circulating in France before. Thus, a memorandum by the Canadian Deputy Director of Medical Services for the Canadian Corps states that in March and April "the prevailing illness was influenza. An epidemic originated from billeting too many troops at one time in a cave. The three battalions crowded there have suffered from many cases of influenza. Fortunately the illness is of short duration and few have been evacuated on account of it. "

Similar reports were received from the medical officers of the 1st and 2nd Canadian divisions, which even suggested that some units were incapacitated because of the disease. "This influenza runs a very rapid course," wrote the 1st Division's Assistant Director of Medical Services (p. 79):

> A man being prostrated for 36 to 72 hours, sudden onset with a temperature from 102-104 degrees, general malaise and severe headache; small percentage of cases showing marked gastro-intestinal symptoms. Three or four day men are able to return to their respective duties. It is apparently extremely contagious –often several cases occurring in a billet within 24 hours time. Reports from the Medical Officers of other units show that there are practically no complications or sequaele.

According to Osborne, censorship would be responsible for not letting out news of the flu in Canada because "Canadian papers were prohibited from printing stories that might give aid or comfort to the enemy or damage morale at home."

Santiago Mata: 1917 American Pandemic

Transfer to Europe: Spain, the scapegoat

To face the German offensive unleashed on the western front on March 21, 1918, an order was issued to activate the arrival of US reinforcements to France, because marshal Ferdinand Foch (1851-1929) believed in the need to impress the enemy with a surprising concentration of forces, although these should be sacrificed, according to the first volume of the complete works of the marshal summarized by Martin Motte (p. 9):

> Joseph de Maistre affirmed that "a lost battle is a battle that one believes to have lost". Foch replies: "A won battle is a battle in which one does not want to recognize oneself defeated". Foch does not let himself be carried away by voluntarism, that is, by the illusion according to which the moral forces alone would decide the battles. He specifies, in effect, that one should not "advance at any price". But the decision does result from the "moral superiority" of one protagonist and the "moral depression" of the other. Therefore it is necessary to aim at the morale of the enemy in the first place. "The means of breaking it, of demonstrating to the adversary that his cause is lost, is the surprise that, contributing to the struggle with something unexpected and terrible, takes away from the enemy the possibility of reflecting and consequently of arguing".
>
> To illustrate the theme as a whole, Foch takes the example of the battle of Wagram (1809), at the end of which Napoleon formed a huge column of 22,500 men and launched it directly towards the Austrian front. Tactically, the measure was aberrant, because such a column offered a huge target for the adverse fire and could hardly return it, since only the first rows could use their rifles. However, it showed a determination that left the Austrians stunned. That was the surprise effect that Napoleon was looking for. When arriving, the

column counted only with 1,500 men, having remained the others dead, wounded or simulating
to be so. But the enemy had fled. In short, "the
troop decimated" had beaten "the decimating
troop", proving that the victory was "purely moral
action."

Whether or not were the Germans as easy to impress in 1918 as following
Foch had been the Austrians a century before, the transportation of so many
American soldiers required not only the support of the British fleet, and of
the Dutch merchant fleet requisitioned, but the permission of the military
doctors, who do not seem to have objected to the displacements until September 24.

The movement of the North American divisions towards Europe preceded,
accompanied and followed the outbreak of the pandemic flu of 1917-1918.
The units that formed the AEF and its displacement order were as follows:

1st Infantry Division: Embarked from June 14 to December 22, 1917. It could
not import the flu, but could get it in contact with other US troops in France,
and transmit it to allies or Germans, since it relieved two French divisions on
April 25 in the Montdidier sector.

2nd Division: In France on October 26, 1917, was able to do the same role as
the previous one in the southeast of Verdun, where it was stationed with
Frenchmen.

41st division: Supposedly shipped on November 26, 1917, when the 81st brigade (regiments 161 and 162) had to embark. The 82nd, composed of the infantry regiments 163 and 164, boarded the ship *Leviathan* to Liverpool on
December 15.

32nd Division (63rd Infantry Brigade, composed of the 125th and 126th Regiments, 64th Brigade composed of the 127th and 128th Regiments, 57th Field
Artillery Brigade - nicknamed the Iron Brigade - with four regiments: 120th):
the first troops left Camp MacArthur (Waco, Texas) on January 2, 1918 to
board in Hoboken, moving to Camp Merritt.

Transfer to Europe: Spain, the scapegoat

Their first casualties occurred on February 5, 1918 when the German submarine U 77 sank *SS Tuscania* (14,384 gross registered tons, GRT), which had sailed from Hoboken on January 23 under Peter McLean with 2,397 soldiers in the direction to Le Havre in the HX-20 convoy. In addition to the majority of the British crew (despite being an American ship), 2,187 soldiers were rescued, so that at most 210 died. Of the two million American soldiers that would be transported between May 1917 and November 1918, 637 (0.03%) died in the sea victims of submarines.

The 32nd division was assigned the role of warehouse or depot from which troops would be taken as needed by other divisions, and which in turn would be relieved by newly arrived soldiers until division was assigned. But finally its chief, Major General Hann, persuaded Pershing to let them fight as a specific unit and fought in the 10th French Army of General Charles Mangin. On May 18, the division relieved the French troops in Upper Alsace in the 27 km from Aspach-le-Bas (west of Mulhouse) to the Swiss border, being the first US unit to set foot on German soil.

The 3rd Infantry Division, which in July 1918 would earn the nickname "Rock of the Marne " was formed on November 21, 1917 in Camp Greene, where as we saw had 565 cases of influenza in December 1917, so it could be the first unit to suffer the pandemic. His chief Joseph T. Dickman embarked on the *Leviathan* on March 4, 1918 and the first units did so on April 4, arriving in France on May 30, standing the next day between Chateau-Thierry and Hill 204, and crashing on July 15 with the Germans who crossed the Marne.

The 4th division was also constituted in Camp Greene, on December 10, 1917, leaving the camp where it could become infected with pandemic influenza on April 18, 1918, bound for Camp Merritt and Camp Mills and embarked on May 1, to train along with the British from June 3 and from the 15 of that month in La Ferté, at the disposal of the 164th French division. They participated in combats in Haute-Vesnes, Courchamps, Chevillon, St. Gengoulph and Sommelans until July 22, when they went to reserve.

As for the 5th division, coming from Camp Logan (Texas), it passed at the be-

Santiago Mata: 1917 American Pandemic

ginning of March 1918 to Camp Merritt and its 9th and 10th brigades embarked on the *Leviathan* to Liverpool on March 4, 1918. It was completed in El Havre (France) on May 1, 1918. From there it was sent to train in Bar-sur-Aube and on June 1 was sent by train to the Vosges, joining the 21st French division in the sector of Colmar until July 16.

The 82nd division left Camp Gordon on April 9 and, through Camp Upton, embarked from Boston and Brooklyn from April 25 to Liverpool, except for its artillery, which went directly to France, arriving on June 1 its last units to Le Havre and training with the British in Escarbotin, west of Abbeville, until June 16.

The 77th Division (*New York's own*) left Camp Upton on March 28, 1918. Some of its units boarded on April 24 the *Leviathan*. Except for the artillery, its units disembarked in Liverpool, crossed England and returned to embark to finish in Calais, whereas the artillery left directly from New York to Brest. It was trained by the 39th British Division in St. Omer until June 16, when it traveled by train to the Baccarat sector.

The 33rd Division left Camp Logan on April 23 to Camp Merritt: On May 22, 1918, it embarked on *Leviathan*, from Hoboken to Brest. The last units arrived in France on June 11. It passed through Huppy, in the sector of Abbeville, and on June 21 was in the sector of Amiens, remaining there until July 4.

The 80th division left Camp Lee on May 17 to embark five days later aboard the *Leviathan*. Its ports of destination were St. Nazaire, Bordeaux and Brest, regrouping in Calais, from where at the beginning of June it left to train in Samur with the British. On June 19, its last units arrived in France and on July 4 it left Samur in the direction of the British Third Army sector, arriving the following day.

The 6th Division came from Camp Forest (Georgia) and Camp McClellan (Alabama), but since May 8, 1918 it was established at Camp Wadsworth, embarking its first units in New York that same day and disembarking in England from July 17 (although the engineers arrived in Brest on May 18 and were

Transfer to Europe: Spain, the scapegoat

employed in constructions in Gievres), moving to Le Havre three days later. On August 28, its last units arrived in France.

The 37th Division was formed at Camp Sheridan (Montgomery, Alabama), from where on May 20, 1918 it was sent to Camp Lee (Virginia) without artillery, passing on June 11 to Hoboken and embarking three days later aboard the *Leviathan*. It arrived in France on June 22, while other units (the 74th Infantry Brigade and Engineers) left Camp Lee on June 21 and arrived in France on July 5. The artillery and sanitary train left Camp Sheridan on June 14, went through Camp Upton and embarked on June 27 to England. The infantry was trained in the Bourmont area and on August 4 it moved to the front in the sector of Baccarat with the 6th French corps.

The 79th Division trained 80,000 soldiers in Camp Meade, Maryland, transferring the majority to other divisions and retaining 25,000: Embarked in Hoboken on July 8, 1918 in the *Leviathan* and landed in Brest, except for the artillery, which embarked in Philadelphia to England (and only after the armistice met his division), its last units arrived in France on August 3.

The 7th Division was trained at Camp MacArthur (Waco, Texas) from February 5, 1918, leaving on July 18 for Camp Merritt and embarking from July 31 in Hoboken (two regiments sailed on August 6 on the sixth voyage of the *Leviathan*) and training since August 19 in Ancy-le-Franc. The last units arrived in France on September 3.

The 86th Infantry Division, which was trained at Camp Grant and there suffered the first wave of pandemic influenza, embarked from New York to France in August 1918, when in theory the second wave had not been declared in the United States. However, its commander, General Lyman W.V. Kennon was barred from boarding because of his poor health and died on September 9 at the Cumberland Hotel (today Ameritania, at the corner between Broadway and 54th street) of pandemic flu: from the complications of a first-wave virus supposedly already extinct , or the virus of the second wave, supposedly not yet present in New York?

The 8th Division sent 5,100 men from Camp Fremont (California) in August

Santiago Mata: 1917 American Pandemic

1918 to Siberia. After covering the gaps with new recruits, on October 18 it moved to Camp Mills (Long Island) and embarked on October 30 some of its units in Hoboken to France, where they did not see action.

The 31st Division was trained at Camp Wheeler (Georgia), embarking on September 16, 1918 and arriving its last units in France on November 9. It was sent as reinforcement to Le Mans, relocating its personnel in other divisions.

The 34th Division was trained in Camp Cody (New Mexico) from October 2, 1917 to September 1918, embarking from September 16 to England and reaching its last units France on October 24. It was sent to Le Mans and dismembered.

As non-shipped units, we must highlight: The 9th Division, created on July 18, 1918 in Camp Sheridan, Alabama, and dissolved on February 15, 1919 at Camp Sheridan. The 12th division, organized on July 12, 1918 at Camp Devens, where it was dissolved on January 31, 1919.

It is usually assumed that the name *Spanish flu* derives from the fact that Spain was the first country where the presence of the 1918 pandemic was widely reported. We will have to ask ourselves, therefore, if in the countries where it originated, the epidemic had less virulence or if it was not spoken about it, as some say, because of the war censorship.

The passage from the United States to France and Great Britain is already known to us. Now it will be necessary to see how the flu spread from the French front to the populations of the other European countries, including the two neutrals (Switzerland and Spain) which, in reverse order to how they were infected (because Switzerland was first), made known the existence of the pandemic.

Beatriz Echeverri, in the work that still in 1993 she titled *The Spanish Flu*, accepts for the ascription to Spain of the origin of the pandemic flu of 1918 an argument as innocent as that of supposing that the evils come from the immediately neighboring country:

Transfer to Europe: Spain, the scapegoat

This theory was sustained in many parts of the world and by such prestigious institutions as the Royal Academy of Medicine of Great Britain during the beginning of the epidemic. Thanks to it the pandemic of 1918 is still known by the name of *Spanish Flu* or *the Spanish Lady*.

Such a practice would have been justified, according to this author, in only one case:

> Dr. Ricardo rightly wrote in 1918 in Portugal that "for us the qualification is absolutely accurate, since it is from Spain, our neighbor, from where we receive the present infection. But we are the only nation that uses it properly. " In the case of France, from where the contagion came to Spain, the inappropriate use of that name could have been due to French xenophobia, exacerbated during the war by the continuous transfer of Spanish workers. However, the idea of the Spanish origin of influenza spread beyond its neighbors. The reason seems to have been the neutrality of Spain during the war. According to Kaplan and Webster "Spain obtained this dubious distinction as a result of the censorship imposed by the Allied and German troops and the little desire of the authorities to recognize such a wide incapacity among their troops." The Spanish press, on the other hand, gave broad coverage to the news about the epidemic wave of spring from its beginning.

And yet, it is possible that what really happened was the opposite, that among those who imported the flu to Spain was a Portuguese, the former president Bernardino Luís Machado Guimarães, who was deposed by a coup d'état, and on December 15, 1917 travelled via Madrid to Paris, where he arrived a month later, taking up residence in the French capital until April

Santiago Mata: 1917 American Pandemic

1918, when he left for Hendaye. On May 14, 1918, the newspaper *El Noticiero de Cáceres* reported on his page 2 that Machado had been sick with the flu and then had met with Portuguese politicians and at least one Spaniard:

> From San Sebastian was reported that the former president of the Portuguese Republic, Don Bernardino Machado, arrived in Hendaye, from Irún, still convalescing from the flu that he has suffered during a season.
>
> There he met with the former Portuguese Prime Minister Alfonso Costa and several politicians and military compatriots, whom he presented with a banquet, including the former mayor of San Sebastian, Mr. Laffite.

By then, according to Françoise Bouron, there already existed in France a report written on May 12 by a doctor about the outbreak of flu recorded in the Fèrebrianges automobile training camp, in the Marne:

> On April 30, 23 indigenous people presented themselves for review, complaining of morbid symptoms that suddenly appeared in the preceding hours: fever of 38 to 40 degrees, headaches, curvatures, congestion of the face of the conjunctiva and pharynx. The convalescence is long and the patients present new symptoms: cough more and more imperious, pneumonia. Persistence of congestion gusts. Within the epidemiological conditions, only the flu, or almost only it, could be taken into consideration.

The epidemic in that French town ended on May 11, having suffered it 30% Europeans and 80% of Indochinese, two of whom died. In mid-May, according to the research published by this author, the percentage of soldiers suffering from influenza in French armies varies between 10, 50 and up to 75%. But the French press does not publish anything because, in addition to the German offensives, the aerial bombings of Paris since the end of January and

Transfer to Europe: Spain, the scapegoat

the *Big Bertha* gunshots since March "make it understandable that, under these conditions, the *Spanish flu* is not the first concern of the French ... and the journalists. "

According to this author, the censorship could not be the cause of not publishing anything about the epidemic in France, since until mid-June only one information concerning it-precisely the case of the Annamites of Fèrebrianges-was censored:

> This silence is due to the fact that the flu is a disease that does not cause fear, as would cholera, typhus or even the plague. It is very possible.

Anyway, by the time French doctors drafted their best-known report on the epidemic in the ranks of their army, the Swiss army had passed the worst of the disease, always in the soft version of the first wave. This is stated in the article on "The flu in 1918 in the 1st division", signed by G. Audeoud and published in 1923 by the *Revue Militaire Suisse*.

According to this account, on May 7, five sick soldiers were evacuated and on the 10th, a record number of 39 fell in the 4th Infantry Regiment, which guarded the border with France. For the company IV-10 it is specified that it was contaminated on May 20 in Tramelan (in front of the French border in the area south of Belfort), with 24 soldiers evacuated in two days.

Relieved the 4th regiment by No. 38, this suffers a flu peak from June 17 to 20, from 20 to 91 patients. The epidemic decreases since June 28 in that regiment. For its part, the Mountain Infantry Regiment No. 5 became rapidly ill on June 1 in St-Imier, in the immediate proximity to the French border also in the canton of Jura: on June 3 it had 61 patients and on June 13, 112.

Taking data from the French authors Dopter, Bassères and Voivenel, the study of the Swiss army recalls that only in the month of May the French army had counted 25,400 patients with flu and 24 deaths, but that these would be 24,282 patients in September and 2,124 deaths, - growing in October to 75,719 and 6,017, which means a total of victims of both waves in the French army in 200,000 sick and 12,000 dead.

Santiago Mata: 1917 American Pandemic

In the Swiss army, the flu reappeared on July 6 in Biena (canton of Bern, at the northern end of the lake of the same name, and south of the Jura) with 9 patients, who would be 72 on the 12th. Between mid-July and in early August, the military hospital in that town had around 250 patients daily. The mountain infantry regiment No. 6, mobilized on June 24, had 14 patients on the 26th and on July 13 it reached a record of 953, being immobilized since July 4 in Freiburg, after passing Château-d'Oex (in the Gruyère regional park) and without reaching its destination in the Jura.

The Freiburg infirmary was closed on August 7. The balance of the first wave in the 1st Swiss division was 82 dead, which represented 2.5% of the total number of patients (3,280) and 1.10% of the troops (7,454, which incidentally implies that 44% of the soldiers fell ill).

As for the civilian population, the epidemic affected more than 50% of the population of Geneva, according to C.E. Ammon, who affirms that "the first cases were reported in 1918 at the country's frontiers, and in foreign soldiers camps, then in inland villages (Château-d'Oex)", for which he cites newspapers from the 5th and 7th July, therefore two months after the first cases declared in the Swiss army. The Swiss Historical Dictionary estimates those affected at two million and the deaths between July 1918 and June 1919 at 24,449 (0.62% of the population, which would therefore be 3.9 million).

In Spain, according to Echeverri, it was on May 22, 1918 when the front page of the newspaper *ABC* dealt with the flu for the first time, "although it is likely that flu cases had been occurring since the first days of the month":

> The festivities of San Isidro, with the agglomerations typical of the festivals, dances and bullfights, promoted the spread of the virus. It seems that in the capital some passers-by were so suddenly attacked that they fell into the street, but the press was quick to reassure public opinion: the "prevailing disease" was not serious, in the Provincial Hospital there was not a single hospitalized case. In reality, it was a nasty infection that

Transfer to Europe: Spain, the scapegoat

was reduced to "two or three days of sagging arms and a limp body."

The news, however, did not appear on the cover, but on page 16, and from what is written, it is evident that the flu was already an epidemic then:

> A rare thing, indeed, it turns out to find a relative, testamentary or friend who is not sick with the flu or who is convalescing from it. This ailment has taken hold of us in Madrid, and not as a peaceful *isidro*, but as an annoying guest. In the asylums, in the barracks, in the tenement houses. In case we had little to scratch, that infirmary for new exciting.

Whoever did headline "The reigning disease" one of his information from Madrid was *La Correspondencia de España* on page 3 of that day 22, with the pretitle of "No need to be alarmed":

> We had been silent until now to prevent the itch of amplification that invades everything from coming to alarm the Madrid neighborhood, fed up with other concerns and demands of today's life.

> The epidemic disease, which fortunately is not serious, although it is very annoying, has invaded thousands of people, and there is no Madrid house, no official center, no theater, no factory, or workshop that does not have numerous casualties as a result of this disease, which deserves the qualification of mysterious, since it was hitherto unknown, and whose characters are not presented frankly, although it bears a lot of resemblance to the flu.

> Its symptoms are vomiting, fevers, diarrhea, and pain throughout the body.

Santiago Mata: 1917 American Pandemic

The agents that produce this disease have not yet been seriously determined. Some men of science are already concerned with this matter, and several seem to affirm that the cause that determines the disease may be the removal of the soil and subsoil due to the major sewerage and other similar works that are being carried out in Madrid.

The phenomenon is more intense and of longer duration when the earth removal is done in days of high temperature.

However, these assertions have no serious basis, since no specific diagnosis is known so far.

Dr. D. José Pin, from the Casa de Socorro de la Latina, made the following statements to a journalist:

"The infirmary has increased a lot these days, to the point that the eight outings that the doctors of said establishment used to make, the day before yesterday and yesterday, have risen to an average of twenty.

The symptoms that patients offer are those of a strong flu attack.

Characters of extreme gravity: it does not present any so far.

Some have thoracic and other intestinal manifestations.

People should take care of themselves, refrain from eating raw fruits, vegetables and legumes, salads and, ultimately, all those foods that can cause stomach upsets. "

According to the conclusions of the thesis prepared by María Isabel Porras

Transfer to Europe: Spain, the scapegoat

Gallo in 1994 (pages 651-667), the pandemic occurred in the capital of Spain in three outbreaks: the first from mid-May to July 6, with the peak being on 27 from May to June 9; the second from the first days of September to December 13, with greater intensity from October 20 to November 16; and the third from mid-February 1919 to mid-May, with greater severity from February 25 to March 22.

The increase in mortality in the capital, according to this author, was "important" but "lower than in other parts of our country", highlighting the first outbreak in June. Mortality from other respiratory processes also increased, with the exception of acute bronchitis, "which was even lower than that of 1915", and which would rise again with the new influenza epidemic, no longer pandemic, of December 1919 and January of 1920, especially in children under one year of age. She would also highlight that "the first outbreak in the city of Madrid was not as benign as the different contemporary testimonies indicated".

As elsewhere, in Madrid young people —individuals aged 20 to 39— were the main victims: "more than 40% of all deaths in 1918 and a little less in 1919", while infant mortality from influenza decreased, and it rose again in the winter of 1919 to 1920.

Once the fear of talking about the flu was lost, on May 23 at least four newspapers covered the topic: *El Debate*, which on page 2 titled "Epidemic in Madrid and Barcelona":

> The epidemic that began in Madrid for two or three days is growing in an astonishing way. Thousands have already been attacked by the strange, benign and short-lived disease, which is manifested by fevers, vomiting, diarrhea and severe pain throughout the body. It bears a strong resemblance to the flu.

After that, the newspaper attached a note from the City Council of the capital that speaks of a "disease of very short duration, which does not offer any seriousness":

Santiago Mata: 1917 American Pandemic

> From the investigations carried out, it seems to
> be a catarrhal infection of extraordinary power of
> diffusion, caused by the prevailing weather con-
> ditions and aided in certain centers, such as thea-
> ters, cafes, barracks, etc., by the agglomeration
> and permanent contact of people.

The note assured that in the Provincial Hospital "there is not a single case of the aforementioned ailment" and that in the barracks "in which there have been numerous invasions, the sick are not transferred to the Military Hospital." The bacteriological examination would show "the presence of germs of catarrhal diseases, with the absence of the germ that causes flu infections" and it was ensured that the waters of Madrid "are in normal conditions (...), nor should the prevailing disease be attributed to the removal of land ", which El Debate corroborates with a telegram that arrived "at dawn " indicating that the epidemic had also been going on in Barcelona for days:

> BARCELONA 22.– The flu epidemic continues. In
> some shops and offices almost all the depend-
> ency is ill. Fortunately, in no case have there been
> serious consequences. In the Alba de Tormes
> battalion almost all the soldiers are attacked.

On the same date, the Soria newspaper *El Porvenir Castellano* published on page 2 an article signed on the 22nd by Eduardo Andicoberry (Erandio, 1888-Tetuán, 1948), joking about the epidemic:

> Half of Madrid, if not all, is these days wrapped up
> in blankets, sweating the fat drop, to cure a "cold
> epidemic". Perhaps the Teutonic espionage has
> influenced this catarrhal streak, which exudes a
> polar freshness, which the most celebrated char-
> acters of Arniches would like for themselves.
> Luckily we have something to thank the flu. Be-
> cause the numerous casualties produced in the
> barracks free us from listening to the "Soldier's
> Song" at all times, and those produced in the ar-
> tistic casts, from having our taste ravaged by

Transfer to Europe: Spain, the scapegoat

wearing cranky calves and flaccid udders. Let's say nothing about the aborted speeches in Parliament, the lyrical chronicles that remain in the pipeline and a thousand other advantages that the "epidemic" will provide us.

Also on the 23rd, on page 2, *El Progreso* (Lugo) wrote that "it has been confirmed that the disease called influenza has appeared in this [royal] Court as an epidemic. In theaters and centers where many people gather, it is observed that hundreds of people are prey to this disease ".

For its part, *La Correspondencia* repeats on page 3 the same text that it published on the 22nd, adding the note from the City Council and another from the "Undersecretary of the Interior" to whom "the Inspector General of Health" (Manuel Martín Salazar), after meeting with "various medical eminences" he would have confirmed that the "said epidemic" had "all the symptoms of the so-called vulgarly *trancazo*":

> It looks like a benign flu infection, whose microbiological study is being finalized by several doctors. It is widely spread and contagious; for there are cases in which it includes a whole family, even the servants. It begins with high fevers up to 40 degrees, some headache, broken bones, and some throat symptoms. (...) The means of preserving against the said epidemic are the isolation of the patient, the disinfection of the rooms and a lot of airing. The spread is greatest in places where many people gather.

Finally, *La Prensa*, from Santa Cruz de Tenerife, on page 3, notes at the same time the spread of the disease and the decrease in alarm:

> A great flu epidemic has developed in Madrid, having acquired a considerable increase in the last 24 hours. The number of attacked is counted by the thousands. Among them are the Supply Commissioner Mr. Ventosa, and other important

Santiago Mata: 1917 American Pandemic

> political figures. (...) The great alarm that occurred in the first moments has partly disappeared because it has been proven that the disease is short-lived.

On May 24, four newspapers dealt with the epidemic again, stating *El Debate* on page 4 that it had spread to La Coruña, where "several employees of public offices have been attacked. At the Central Post Office, an officer and nine postmen are sick".

However, *El Eco de Santiago* (page 1) only talks about an epidemic in Madrid and although it repeats the statement of its low severity because "not a single death has been registered", it already recognizes that "nursing is increasing and is a cause of constant concern for doctors, who have been doing work these days beyond their strength, if one takes into account the extraordinary number of visits they are forced to make every 24 hours". This newspaper repeats the hypothesis about a cause derived from the Metro works, citing a case where the blame also ended up falling on Spain:

> The not distant case of a similar phenomenon that occurred in Caracas on the occasion of the underground works is remembered. The epidemic was of such a nature that the North American Government sent a scientific Commission, and it diagnosed that it was the reappearance of microbes of a disease that must have existed no less than in times of Spanish domination.

The same hoax about the subsoil appears on page 2 of the *Diario de Reus*, which limits to Madrid what it calls "general epidemic state that lasts three or four days," but then that newspaper copies from the *Diario de Tortosa* a news item not related to the flu in the capital:

> Many cases of acute colds, flu and gastro-intestinal are noted in this city. These diseases, typical of the current season on a more or less scale, do not appear regularly with alarming symptoms, being able to combat them in three or four days.

Transfer to Europe: Spain, the scapegoat

For its part, *La Rioja* (page 2, news dated 23) reports that fear is spreading in Madrid - and convalescence in bed is already prolonged to six days -, in Barcelona many patients are admitted to the military hospital and there is an epidemic situation in Sevilla:

> Among public employees, the prevailing illness is noticeable in an extraordinary way, since there are offices where no more than two or three of the many who provide service have attended today. Although the epidemic has not acquired serious characters so far, fortunately, the alarm spreads among the people, who are cowed, noticing the effects in public walks and shows, cafes and other places, usually crowded. (...) The disease does not present serious characteristics, and lasts from four to six days, during which it is necessary to stay in bed as a result of the laxity that takes over the organism.

> A music band that had to go to the replacement of the Palace on Tuesday [May 21], could not attend it because half of the musicians were ill. The reigning catarrhal epidemic reached Congress on Wednesday [22]. Two journalists who made information were suddenly attacked by the strange evil. One of them needed help in the Chamber medicine cabinet. The Barcelona and Seville newspapers report that the epidemic is also spreading in those capitals. In Barcelona, many attacked are admitted to the Military Hospital. So far, the epidemic has benign characters, and no deaths have occurred as a result of it.

In France, *Le Figaro* dedicates an article on page 2 to "The offensive and the influenza", and begins by echoing alleged riots in the German army as a cause of "the inertia of the German infantry":

> Another rumor, quite curious, but confirmed, comes from medical media. *It would seem that a*

> *very strong influenza epidemic is ravaging the enemy army...* Today we can set the number of effective combatants of that army on our front. The number is 1,450,000 men. Adding to these effective combatants those of the quartermaster, supplies, ammunition personnel, etc., etc., we arrive at a figure of approximately two million men. *(Radio.)*

On May 25, *El Adelanto* notes on page 1 that the disease "we have also suffered from it in Salamanca for a few days, although, fortunately, with the same mild characters with which the epidemic has developed in the capital of the kingdom", although they dare to disagree in the diagnosis:

> The characteristics of the disease, which does not seem to be the flu, although clinically it is very similar to this, are colds of the upper airways, high fever, dejection and digestive system disorders. It seems that the way the patients become infected is not due to food, but is due to the air breathed, this being due to the great atmospheric changes suffered, and the spread of the ailment is, in many cases, by contagion.

El Debate, on page 4, adds Palencia to the list of affected cities and for the first time gives a number of soldiers admitted to hospitals in Barcelona - while warning that the epidemic is spreading to garrisons throughout Catalonia - quoting the mayor of the city, according to whom "the number of attacked is enormous":

> In a store in the Plaza de Santa Ana a sign has been posted that says: "Closed because the entire dependency is sick." In the Military Hospital, of the 562 troops who receive assistance there, 155 are attacked by the unknown disease. In the garrison regiments of all the Catalan provinces the number of invasions is equally high.

Transfer to Europe: Spain, the scapegoat

In Palencia, this newspaper indicates "several cases of a strange disease. It is supposed to be the same flu epidemic that exists in Madrid and Barcelona". In La Coruña, "flu cases are increasing. Upon the arrival of the mail, the mail wagon was disinfected".

The same newspaper indicates on the 26th on its page 3 that the epidemic exists in Burgos and possibly Murcia:

> Burgos 25.– The prevailing disease in other towns has occurred in this city, with a very large number of those attacked, especially among the soldiers. There are 200 patients in the San Martín Infantry regiment. At the Bourbon Cavalry Barracks an entire squad has been attacked. Once the guard service was carried out in the prison and in the powder magazine, it was necessary to relieve nine soldiers who felt ill. In the offices, many employees have been terminated due to sickness. The disease is benign.

> Murcia 25.– There have been several cases of an unknown disease in this capital and in some towns in the province. Doctors don't think it's the flu.

La Rioja of May 26, on page 2, reports, in an information dated 25 in Madrid, that "the ministers of Public Instruction and the Navy were absent from Congress because they were ill with the flu."

On May 27, 1918, *ABC* published a note from the Provincial Board of Health chaired by the civil governor Luis López Ballesteros, in order to "stop, as far as possible, the prevailing disease", which was defined as "flu-like in nature. ", Ruling out that" it can be attributed to contamination of the waters from which Madrid is supplied or to the removal of subsoil lands due to the Metropolitan works, or other similar ones ". It was stated that "the germ of the disease swarms in the atmospheric air" and its proportion depended "on this air, depending on whether it is more or less confined". He erred in considering it not very deadly:

Santiago Mata: 1917 American Pandemic

> This disease has great diffusive power and little virulence of the microorganism that originates and spreads it. This is, without a doubt, a reason for the neighborhood of this capital not to be alarmed, without for this reason ceasing to recommend the greatest care in the hygienic and dietary regimen of the sick and those who care for them, as it must be remembered, among others, the flu epidemic of 89, which began in Madrid in a benign way and ended up taking on serious and fatal forms in many cases, not only due to the nature of the process, but also due to the complications and consequences of convalescence.

As for concrete measures to "stop its development as much as possible" there was talk of "healthy eating in all senses", although it was stated that neither food nor water could be "carriers of the germ". It appeared later, but its importance was evident from the tone, the "convenience of not breathing the air of atmospheres confined to cafes, taverns, public shows, casinos and other places of agglomeration", advising "the oxygenation of the lung" through "walks to the open air, and better in the field, and sun of the organism; oxygen and light, disinfectants par excellence for the lung and skin ."

It ended up advising as "essential" to clean the clothes of the sick and related to them, as convenient to ventilate their rooms, as "useful to isolate the healthy from the sick as much as possible", assuming that the heat favored " the development and life of the germ "in the respiratory system, so it would be" more dangerous for healthy people to breathe in the atmosphere that surrounds a living patient than that surrounding a deceased person."

On the other hand, the purity of the water was not useful nor was there known "any prophylactic medication to avoid the presentation of the process or to stop its development, nor are there any known serums or vaccines that can preserve the disease or help us fight it. The only effective preservative is based on the possible isolation of the healthy and the sick, ensuring that the former remain in the atmosphere that the latter breathe only for the time necessary to attend to their care".

Transfer to Europe: Spain, the scapegoat

La Prensa (Santa Cruz de Tenerife) of Monday, May 27, page 1, refers to the convalescence due to influenza of the ministers Alba (Instruction) and Pidal (Navy), of the president of the Congress, Villanueva, and reports that there are already dead: "In Valladolid, the director of that Institute, Mr. Policarpo Mingote, has died as a result of the flu" (Policarpo Mingote y Tarazona, born in 1848 in Granada, director of the Secondary Education School located in the Plaza de San Pablo, today IES José Zorrilla). In Madrid, "the composer Don Luis Foglietti, who was attacked with pneumonia," died.

On the same day, the weekly *La Defensa*, from Sigüenza, reports (p. 3) the presence of the epidemic there:

> Until today some cases are known in this city and in the immediate town of Moratilla de Henares, of the disease that reigns so much in Madrid and that according to the doctors is only "a benign flu".

In Melilla, *El Telegrama del Rif* reported (p. 2) that on the 25th (since the news is dated 26) the first cases occurred in Ciudad Real:

> In this capital the flu epidemic has occurred, with the same intensity of character as in Madrid. In extreme neighborhoods, since yesterday, when the first cases were reported, these increase in an extraordinary way, with entire streets in which all their neighbors are attacked by this strange disease.

El Debate, on its cover of the 27th, has an article signed by Armando Guerra (pseudonym of Lieutenant Colonel Francisco Martín Lorente) on the military situation in which he affirms that the flu is defeating the Germans, as published on May 24 by *Le Figaro*, and quotes another newspaper that did not want to wait for the US deployment to go on the offensive:

> I don't suppose the allies will wait for the new fighters to come online to attack. No, the *Petit Parisien* responds: "If the enemy persists in his

Santiago Mata: 1917 American Pandemic

immobility, we are the ones who will attack." It's okay. Either you attack, or I attack, because the Germans, according to *Le Figaro*, IS PROVED that they have been attacked by the flu microbe, and it must not be difficult to defeat them, when they have surrendered to a microscopic enemy. Despite this proven news and despite the *Petit Parisien* arrests, the truth is that the lines are not moving an inch.

On May 28, *La Prensa* of Santa Cruz de Tenerife (p. 3) reported that they were "attacked by the reigning disease, King Alfonso" and political personalities such as "Mr. Dato, Argente, the Undersecretary of Finance, and numerous deputies ".

El Debate, on page 4, also reports that the monarch has been in bed since the previous day, and publishes the City Council's bacteriological examination note. It mentions the visit of López Ballesteros to inspect the Provincial Hospital where "few are the patients who are sheltered there who suffer from the current disease, and entry into the Hospital varies very little from normal." It specifies four new cities victims of the epidemic according to information dated 27, starting with Cartagena:

The flu has become widespread in this city, with many attacked in the regiments of Seville and Spain, and about a hundred sick in the navy. There are also a large number of attacked in the artillery detachments, among the Arsenal workers and in the House of Mercy, where 14 children suffer from the disease, who have been visited by the mayor and the health authorities.

In Gijón "some people arriving from Madrid have been forced to stay in bed. The function announced to be held at the Bindurra theater has been suspended because the main artists are ill." In San Sebastián "the flu epidemic has been declared." The number of attacked "is very high. Only in the regiment of Sicily there have been 150 cases". Also in Zaragoza "the flu has been

Transfer to Europe: Spain, the scapegoat

declared epidemic. In the Civil Guard barracks there are 15 attacked. In the offices the disease has also been noticed, and quite a few are sick."

The Vitorian newspaper *La Libertad*, on page 2, when reporting on the spread of the epidemic in the capital of Álava, and that "the first ones that have been primed have been with the postmen", headlines the news "The soldier of Naples, in Vitoria ", without explaining why he gives that name, apart from the fact that a girl comes to tell the editor" that her mother - a delivery girl - will not be able to come today because she is with *the soldier of Naples*.

Supposedly because it was catchy, the 1918 flu was given the name of the choir entitled "Soldado de Nápoles", which was part of the zarzuela *La Canción del Olvido* (The Song of Forgetting), which with music by maestro José Serrano was re-released at the Teatro de la Zarzuela in Madrid on March 1, 1918 (the real premiere had been at the Teatro Lírico in Valencia on November 17, 1916).

On page 3 of the *Heraldo de Zamora*, an article signed by Sánchez Ortiz makes the Barcelona epidemic depend on the Madrid epidemic by assuring - without citing any source - that "there are many cases that are all reported from Madrid".

That was also the day of the first mention of the flu in Le Journal on its 3rd page, an epidemic that affected 30% of the population of Madrid, including the king after he attended mass in the palace chapel with a large crowd . Here, too, the "benign" character of the flu is affirmed. Le Figaro publishes similar information on page 2, titled "The flu in Spain", and also mentioning the president of the congress, the ministers of Finance, Navy and Public Instruction, and the undersecretary of the Presidency. It specifies that "according to the offices of the province, approximately a third of the population is affected."

On the cover of *El Eco de Santiago* on May 29, in an article entitled "Verbas - A frebe de moda" (Words – fever in voge), Xavier Montero notes the use in Madrid of that name taken from the aforementioned zarzuela, trusting that music would scare away the disease (original in Galician language):

Santiago Mata: 1917 American Pandemic

> Fever in voge or soldier from Naples is called a disease that has many people in Madrid in bed to-day. If my life came to grandchildren, I would con-tract my little descendants that one year there was an evil that was called the soldier of Naples. Such a strange name was given to him because the bacilli of the flu were excreted full of genus against humans.
>
> It was then when there was the European war. A soldier from Naples who was marching the bat-tles, said goodbye with a doorid song, from his girlfriend. And the music, which the whole world has claimed, domesticates the beasts, has brought down the wrath of the flu microbes.
>
> These wounded the men. But the feeling of the song voted by the soldier, put tears in their eyes and softness in the wild heart. For him the evil was captive. Bad for three or four days nothing more. Here no one is afraid of fever.

On the same May 29, the civil governor of Madrid, Luis López Ballesteros, issued a statement that referred to the increase in cost of medicines due to the "prevailing epidemic", ordering the return to the prices of May 1. In France, this day *La Croix* refers to the Spanish epidemic.

On May 30, *La Prensa* affirms on the front page, dating it the 29th at 2:35 p.m., that "the number of people attacked by the disease prevailing in this Court increases in a very alarming way," providing a hoax about a food origin:

> At present there are more than 120 thousand pa-tients. Most of the doctors who serve in the Aid Homes, are also attacked by the flu and not at all unable to practice the profession. This circum-stance has increased the panic that reigns among the popular classes. Some newspapers advise that newly imported wheats from America be

Transfer to Europe: Spain, the scapegoat

scrupulously analyzed, in case they turn out to be the cause of the alarming epidemic.

The positive note would be that the king and ministers "who were attacked by the disease, are much improved." The same day, the newspaper El Pueblo, in Valencia, published on its page 3 a phoneme sent the night before from Castellón, where "the appearance of the disease in voge has been confirmed, with 60 attacked". Also "among the force of the cruiser Carlos V the flu has appeared. There are 70 attacked, increasing from hour to hour ". In addition, the appearance in Sweden of an epidemic that was not related to the prevailing one in Spain was confirmed:

> An unknown disease, which presents serious characteristics, has developed in Sweden as an epidemic. All the symptoms show dropsy, attributed to nutritional deficiencies. It causes numerous victims, especially among workers.

On the 31st, on page 2 of *El Adelanto de Salamanca*, Pallol's signature echoed the price of medicines and the food hoax in Madrid:

> Now the flu is supposed to originate in American flour, imported not long ago and spread over several regions. He should not have reached Segovia. I eat Segovia bread - the same one that my dear friend the poet Rodao eats - and I'm still healthy, except for my damn nervous system.

More specific was the information that appeared on page 4 of *La Correspondencia*, specifying that "the general director of Communications has ordered that the officials of the branches of this court go to the Central, with said branches being closed with the exception of the Paseo de Recoletos , until the current circumstances cease ".

In Palencia the flu epidemic would have "considerably increased", while in Toledo "it continues to cause casualties. Every day the number of attacked is greater ". In Cádiz, on the 30th, the director of the Military Hospital –which had 400 beds–, Francisco Alberico, and the military governor, Mr. Felíu, took

Santiago Mata: 1917 American Pandemic

measures "on the occasion of the declaration of the prevailing epidemic in the garrison", since " between yesterday and today more than a hundred attacked soldiers have entered. There are also two doctors suffering from the same thing ".

The same newspaper, after reporting on page 7 of the first two deaths in Madrid, published the number of deaths in the last ten days, presented by the Charity Councilor. Without saying it, they were an accusation of lying for the previous statement, since from May 21 to 30 they went from 46 to 107 deaths (starting the rise from 43 dead on the 26th to 94 on 27, 75 on 28 and 94 on 29).

On June 1, the French Catholic newspaper *La Croix* reported the number of 120,000 patients in Madrid. On the other hand, *La Libertad* of Vitoria raised the figure on its page 2 with the headline "More than 300,000 affected by influenza", providing the data of daily mortality, which "in Madrid fluctuated between 30 and 40, and on Thursday the deceased amounted to 104 and yesterday the death toll rose to 107 ". At the Toledo Infantry Academy there would be "more than 200 students attacked by this disease." Among the new patients in Madrid would be "the Minister of Grace and Justice, Count of Romanones":

> The number of people affected by influenza in this Court is enormous. According to official data, more than 300,000 suffer from the strange ailment. The bakers state that the sale of bread in Madrid has decreased by more than half. On the other hand, the sale of milk amounts to more than double what was sold in normal times.

On its page 3, *La Prensa* already supposed the king "recovered from the epidemic flu disease that affected him. Today he dispatched with the head of the government, Mr. Maura, "according to a note dated 31 at 4:00 p.m. Although the new patients were joined by deputies like Marcelino Domingo, in general the opposite was affirmed than the previous newspaper:

Transfer to Europe: Spain, the scapegoat

> The number of people attacked by the influenza
> epidemic decreases, showing that it itself tends
> to decrease. Almost all those attacked at the be-
> ginning of the epidemic are recovered.

On June 2, *El Debate* reported on page 2 the session held at the Royal Academy of Medicine on the "origin and diagnosis" of the flu, in which Dr. Huertas
said that "it is totally impossible to find out the number of attacked, and that
during the current week mortality has increased somewhat, due to the complications of the prevailing epidemic with kidney and heart injuries, for which
reason it is estimated that the current epidemic is *currently* benign.

Dr. Hernández Briz declared that "it is enough to know the symptoms to di-
agnose the current epidemic, which he considers to be the flu"; and even
more categorical was Dr. Marañón when stating "that, without a doubt, this
epidemic can be classified as influenza." On the contrary, Dr. Pittaluga "is in-
clined to believe that it is not influenza, since the analysis made is missing
the Pfeiffer-producing bacillus, and rather it is another infection."

Meanwhile, the newspaper cited Bilbao as places with an epidemic, where
on the 31st "the reigning disease continues, spreading through the population"; Guadalajara, where on day 1 "in the College of Orphans there are 60
attacked by the epidemic and in the Academy of Engineers, 50"; and Pamplona, where on date 1 "the epidemic in voge has occurred in this population,
with many being sick. Instead, Tarragona still seemed safe that same day,
when the Provincial Board of Health agreed "that every three days public
premises be disinfected, in order to prevent the population from the prevailing disease."

La Correspondencia, on page 2, specified for the case of Pamplona (dating
the events to day 1) that the epidemic attacked "numerous soldiers of the
garrison" and "among the countrymen there are also several cases." In Cádiz
"the epidemic has begun to drop in the Provincial Hospice, where there are
445 sheltered of both sexes. The Military Hospital is full of sick people, and
many others attacked are assisted in the Santa Elena and San Roque barracks", with "numerous cases" among shipyard workers, Matagorda and

Santiago Mata: 1917 American Pandemic

public offices. In Zaragoza, it was "calculate the number of people attacked at 3,000", with 72 cases in the Hospice, 70 in the King's Lancers Regiment, 15 in the Post Office and as many in the urban Police. In Vigo, there were in the San Simón lazaretto "168 crew members of the second naval division, 20 having been discharged and 26 admitted. None are seriously ill and most are convalescing."

On June 6, 1918, the French press echoed the decline in the flu epidemic; This is the case of *Le Figaro* on page 2, quoting information dated the 5th in Madrid:

> The flu epidemic has decreased in Madrid. Its decrease is also noted in provinces, with the exception of those in the northwest. The bacillus whose presence has been found most frequently is influenza, but meningitis has also been recognized. The violent alternations in the climate of the peninsula have evidently favored the spread and violence of the disease.

In this way, references to the first wave of influenza in Spain disappear in that newspaper. In fact, the next news about its southern neighbors will be, on June 13, a news broadcast from Madrid on the 12th about an article in the *ABC*, a newspaper "whose Germanophile tendencies were known", and where France's heroism was now being praised.

The flu would still claim an illustrious victim in Madrid on June 19, in the person of Julián Juderías, a young historian who had tried to wash the image of Spain from the various black legends imposed from abroad, and whom fate now saved from suffering from contemplating how the *sanbenito* of having originated the deadliest pandemic in history fell on his country, as Luis Español recounts on page 61 of his biography:

> The foreign press called that illness the Spanish flu or influenza, which is a way of bringing the Black Legend to the epidemiological field. It is still curious that Juderías died as a result of a black

Transfer to Europe: Spain, the scapegoat

> legend turned into a virus … It was three thirty in
> the afternoon on June 19, 1918 when the Hon. Mr.
> Don Julián Juderías y Loyot ceased to exist in his
> house at 33 Calle de Preciados.

And while Spain stopped being news, England began to be. News of the first wave would pass almost unnoticed elsewhere. According to Debra E. Blakely (p. 23 of *Mass Mediated Disease*), in the United States the first information about the pandemic appeared on June 1 in the *New York Times* talking about China, where there would be a flu-like illness: "The disease is not fatal and happens in the course of four days".

The arrival of the first pandemic wave to China (specifically Canton) was documented by W.W. Cadbury in *China Medical Journal* (1920, pages 1-17). Cheng and Leung state that in Shanghai "it began at the end of May and lasted until June", according to a letter written on February 11, 1919 by Dr. A. Stanley, a health worker in that city, who places the second wave in October and November "with more serious symptoms". These authors estimate that the crude mortality rate from influenza in Shanghai in 1918-19 was 1.3 per thousand, much lower "than the average mortality rates among those infected in other parts of the world in 1918 (2.5% or more)", and even lower compared to San Francisco, where there were 23,639 infected and 2,122 deaths (mortality rate 8.98%).

According to these authors, the flu spread in China from the south (Shanghai and Canton) to the north to Harbin and remote regions such as Gejui, in Yunnan province, with 10,000 people affected in the port of Wenzhou in May 1918, a morbidity higher than 50% in September among students and with the pandemic remitting at the end of October, after having also reached a morbidity of 50% of the population in Beijing, very similar to the 49% registered in the Philippines, where, however, mortality was 2.3%.

A cable dated June 20 in The Hague would have the same newspaper headline: "Spain is affected by the German disease, and other countries will be, according to a Dutchman", who is only said to be a tailor returned from Germany and that attributes the disease to the accumulation of water caused in

Santiago Mata: 1917 American Pandemic

the body by the excess of turnips that the Germans were forced to eat.

We found on June 25 a direct reference to the disease in the German army. But the correspondent for *The New York Times* -Philip Gibbs, who titled his article "Dysentery and Influenza Weaken German Army As Hopes of a West Front Victory Are Waning"- does not specify that it is the same epidemic registered in Spain, nor that it is responsible for weakening the Germans:

> WAR CORRESPONDENT HEADQUARTERS, June 24.- The German lost heavily in a raid carried out this morning by British troops between Fletre and Strazeele. They put up a fierce resistance, and over 100 of then were killed in their trenches in the close fighting, as well as by the British bombardment. Fifty prisoners were brought back and three machine guns.
>
> Other small raids were made by the British elsewhere, resulting in prisoners being brought in.
>
> The enemy is not happy, either with his present conditions in the line or with the prospects of another offensive, which many of the Germans believe they will be called on to make soon. It is obvious that the great hopes they had before March 21 have frittered out, owing to the way their army has been checked and held on all sides. The most optimistic of them still believe they have a last supreme chance of gaining a decision this year, but I doubt whether this belief is widespread among the men now in or near the line.
>
> Their health does not seem of the best just now, and they are said to be suffering to some extent from a kind of influenza, to which they are perhaps rendered more liable by their restricted rations. Dysentery also has touched them, according to information that has reached us. This sick-

Transfer to Europe: Spain, the scapegoat

ness is not, in my opinion, accountable for the delay in launching a new offensive, but may have some effect on the fighting quality of the enemy.

Two days later, the *New York Times* mentions those two aspects that were left out:

LONDON, June 26.- The proportion of men sent to hospital son account of influenza has risen rapidly in all the German units in the last few days, and special hospitals are being established in the rear areas dealing solely with this disease.

Thus far only the more serious cases have been sent to the hospitals, but the German army doctors say that, unless even the light cases are removed from the units, it will be difficult to prevent further spread of the epidemic.

The diseases prevalent in the German Army is reported to be of the new Spanish type, which recently broke out in Berlin and other German cities, and is presumed to have been brought to the trenches by men returning from leave.

In the German cities the disease has been very hard to deal with owing to the shortage of doctors and the conditions of undernutrition among the city populations.

WITH THE BRITISH ARMY IN FRANCE, June 26 (Associated Press.)- German troops on the western front are suffering from an epidemic of grip, which incapacitates them for a week or then days. There are also many cases of typhus and dysentery within the German lines southwest of Lille.

There is no evidence, however, that these illnesses are responsible for the delay of the new

Santiago Mata: 1917 American Pandemic

offensive. It is believed that the German High Command has nearly completed its preparations for the next great attack against the allied front.

On June 28, 1918, an opinion article titled *Concerning "Spanish Influenza"* appears for the first time, although on page 10. Of course, so as not to panic, it is preceded (p. 2) by an article that denied the existence of such a flu in the United States Army, but that, as an *excusatio non petita*, for those who know what happened, explained the effects of the pandemic flu that that army had already suffered:

> NO INFLUENZA IN OUR ARMY. WASHINGTON, June 27.- No advices have reached the War Department about the influenza among the German troops on the western front.
>
> The reported epidemic is not regarded here as having serious proportions. It is clear that the soldier who has it is incapacitated for duty, and thousands may be down with the disease at once, so that military movements may be delayed.
>
> The American troops have at no time shown any form of the disease. Precautions have already been ordered, however, to meet any emergency.

The second of the *Topics of the Times* on page 10 was, as said, *Concerning "Spanish Influenza"*:

> What degree of importance should be ascribed to the report that an epidemic of *Spanish influenza* is sweeping through the German armies on the western front depends on the degree of virulence marking the particular strain of bacilli that is now at work. Sometimes the effects they produce are little more than annoying, but not infrequently they are completely disabling for long periods - several months- and many of the victims die or become invalids for life. It is possible, therefore,

Transfer to Europe: Spain, the scapegoat

that the appearance of this malady may influence the conduct of the war, or even play a part in deciding its conclusion.

I one were allowed to be human rather than humane, there would be a general yielding throughout the allied world to the temptation to express the hope that the Germans may have the disease in a severe form. That would be no more than they deserve, indeed, but for at least one reason -the reason that wishes, in such matters, do not count either way- it is better not to put in definite form a desire which, after all, is worthier of Germans than of civilized people.

Moreover, if the influenza is raging on one side of No Man's Land, it is sure soon to appear on our side of the lines. But that is would be equally disastrous when thus transferred by prisoners or otherwise is by no means certain. On the contrary, there is ground for hoping and even for believing that the soldiers of the Allies, not having been through a long course of undernourishment, would be better able than the Germans to resist the attack of the bacilli, and that their sufferings from such a cause would be comparatively light.

The fact that this influenza is called Spanish by no means proves it of that origin. Dubious, too, is the recently suggested theory that the malady in its present form was started by the conditions produced among men making long cruises in submarines. Yet that theory may be true. Bacilli are as much affected by environment as are any other animals, and the submarine strain of the influenza bacillus might well be measurably unlike its relatives and more injurious to human hosts.

On June 30, *Le Figaro* resumed on page 2 the task of reporting an epidemic

Santiago Mata: 1917 American Pandemic

about which the last news item that appeared in that newspaper had been published on the 6th of that month, stating that it was remitting in Spain. It was already given the name that it will carry for a century and its presence in Germany was affirmed, according to a dispatch dated in Basel on the 29th:

> The "Spanish flu" in Germany. The epidemic disease that struck Spain has suddenly made its appearance in Nuremberg, where large numbers of people have suddenly fallen prey to fever.

On July 2, 1918, it was *The Times* in London who began to speak of deaths from a flu attack in the English capital, and that many people fell ill suddenly, although they recovered after a few days in bed. Dr. R.C. Jewsbury, a Lambeth pathologist, stated: "The prevalent Spanish flu was very rapid in the beginning, acute in its progression, and rapid in disappearing."

On the 3rd, *The Times* referred to Manchester stating that "in terms of numbers, the current one is the worst epidemic in the city in many years", although it insisted on the topics of rapid recovery and soft character. On July 6, it was stated that "the flu epidemic shows no signs of abating." But in mid-July, all information about the flu disappeared *in The Times* until September 12.

Honigsbaum (p. 49-50) states that British military doctors were certain that the route of entry for the "Spanish" flu into Britain would be the soldiers returning home:

> However, for army medics who had seen the flu sweep through the front lines at first hand and had taken note of the extensive morbidity, and for civilian doctors and public health officials who had made a special study of the disease, there were plenty of reasons to be concerned. The first was that although just 5,500 British troops had died from the flu -a paltry figure when set against the dreadful arithmetic of the Somme and Passchendaele- some 226,000 had been hospitalized, a truly staggering number. Moreover, it was these

Transfer to Europe: Spain, the scapegoat

same soldiers, returning to Britain via Portsmouth and other Channel ports, who had introduced the disease to the civilian population, carrying the virus by rail to London and Birmingham from where it was refracted by the train network to northern cities such as Leeds and Liverpool, and west to Bristol and Cardiff.

If, contrary to what has just been stated, the flu had been introduced much earlier by the soldiers of the American Expeditionary Force (AEF), the highest incidence of the first wave would have to be registered not in the Channel ports, but in the localities that surrounded one of the two main ports of AEF landing: Liverpool, where 844,000 AEF soldiers arrived (the other port was the French city of Brest, where 791,500 AEF soldiers landed).

To evaluate the first wave of influenza in England, we have the data published in the aforementioned Newman report. On pages 45-47, it presents the flu death rates per thousand inhabitants, for the three waves of summer, fall and winter. By pointing out on a map the cities in which this value is greater than 2 for the summer wave, we obtain a map in which no Channel port appears, nor London, nor any of the cities mentioned by Honigsbaum, and instead it is consistent with the fact that it was the sick American soldiers who imported the pandemic into the United Kingdom when they landed in Liverpool.

Santiago Mata: 1917 American Pandemic

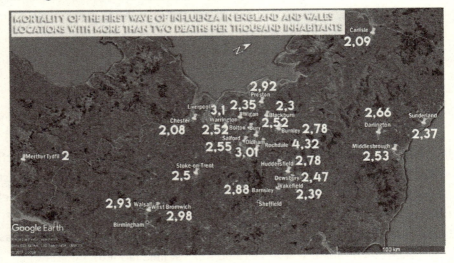

Many of these localities were headquarters of different units of the British Army, which in turn made contact with the American units, although it is very difficult to reconstruct the movements of all the military of both countries to specify which AEF units were the first to spread the flu to the British.

On June 10, according to Killingray and Phillips, the first wave of pandemic flu was noticed in India, when seven sepoy policemen were admitted to the Bombay hospital with fevers other than malaria. One of them worked on the docks of the port and according to the city's sanitary chief, J.A. Turner, the flu was imported by the crew of a ship that had anchored in late May. The government of India, by contrast, will claim that the crew contracted the fever in Bombay.

On the 15th, employees of the shipping company W. & A. Graham & Co. fell ill and on the 20th employees of the shipyard, the Bombay Port Trust, the Hong Kong and Shanghai Banking Corporation, the telegraph, the mint and the Rachel Sassoon factory. That same day a hospital ship (with soldiers) arrived in Karachi and within two days most of the hospital patients had the flu.

From a daily mortality of 92 people in Bombay on June 21, it rose to 230 on July 3, when *The Times of India* reported that "almost every house in Bombay

Transfer to Europe: Spain, the scapegoat

has some of its inhabitants in bed with fever and all offices regret the absence of employees". In that city, the first wave lasted four weeks, killed 1,600 people and was responsible, according to Turner, for the loss of a million days of work.

In Montreal, the military transport *Nagoya* had anchored on July 9, with 100 of its 160 crew members sick with the flu, but until September the epidemic did not affect the city. On July 21, an order had not yet been issued to quarantine the many ships carrying pandemic flu patients. However, Osborne (p. 84) assures:

> An análisis of six military hospitals across Canada demonstrates that despite a clumsy official response, influenza did not spread into the interior during the summer of 1918.

Camp Hill Hospital in Halifax, for example, cared for naval and military personnel as well as those returning from overseas, including the ship *Araguayan*, which arrived in July with dozens of sick soldiers. Still, from July 1 to September 29, the hospital only recorded 285 flu-like cases, 13% of admissions (and an average of 3.13 cases per day).

In mid-July 1918, the flu returned to Québec, aboard the ship *Somali,* quarantined off Grosse Isle, east of the city, nine of whose crew were hospitalized with flu. Within a few days, most of the 177 crew members were hospitalized and by the end of the month the epidemic entered the city, although the authorities baptized it as a pneumonia epidemic.

In *Le Figaro*, the disease reappeared on August 1, 1918, when earlier information about cholera in Saint Petersburg was expanded. Citing the Petrograd *Gazette*, it is said that on July 20 there were 21,200 cases of cholera in the capital, and that it would have spread to Hungary and Sweden, as a report from Stockholm spoke of four cases on July 29, considering the Swedish government "That the danger is considered definitively ruled out" and "passenger transport between Stockholm and Petrograd" had been re-established. On the contrary, in Switzerland, the federal council would have ordered the subject to prophylactic measures "to travelers, luggage and merchandise

Santiago Mata: 1917 American Pandemic

coming from Russia, Hungary or Sweden". *Le Matin* will take over these measures on page 3 on August 4, stating that "especially people from the Balkan countries will have to undergo a thorough examination."

Finally, on August 7, on page 2, *Le Matin* announced that what is in Switzerland is the flu, and for the first time entitled an article "The Spanish flu", although readers could learn from it that that epidemic ravaged also to France, for being in Western Europe:

> It is certainly the flu that has just hit Switzerland. In the course of yesterday's session at the Academy of Medicine, Dr. Jules Renault, a hospital physician in charge of a research mission on influenza, presented his conclusions. It has shown that the epidemic that has just wreaked havoc on the other side of the Jura is an influenza epidemic and that the legends born in Helvetia on account of that disease are absolutely false. Indeed, the investigations carried out in the laboratories of the hospitals and the School of Medicine have shown that the microbe is the cause of the flu and that it is not, as they had said, bronchial diphtheria, nor exanthematic typhus, and even less cholera or plague. Otherwise, the epidemic appears to be in full decline, and according to Dr. Renault there is little reason to fear its spread. Finally, it is identical to the epidemics that recently ravaged Spain and the various countries of Western Europe.

The first wave of pandemic influenza arrived in South America as well, and thus the authors of the study on Peru led by G. Chowell found "evidence of an initial mild pandemic wave between July and September" of 1918 in Lima.

Meanwhile, in the United States, the human pandemic flu of 1918 will give rise to an animal disease that will henceforth be recombined with human disease: the swine flu, which, if we believe what Richard E. Shope claimed in 1936 (page 681 of the article cited in the bibliography), appeared in the late

Transfer to Europe: Spain, the scapegoat

summer of 1918 (a veterinarian named Grant B. Munger observed sick herds in August in western Iowa), evidencing its importance at the Cedar Rapids Pig Fair, Iowa, 365 kilometers northeast of Saint Joseph (Missouri) and 925 of Haskell:

> Conversations with veterinary practitioners in eastern Iowa have revealed that the disease caused serious losses among swine on exhibition at the Cedar Rapids Swine Show held from September 30 to October 5, 1918. At the conclusion of the show, the swine, many of them ill, were returned to their home farms and, within 2 or 3 days of their return, influenza was stated to be rampant in the portion of the drove that had remained at home. Shortly thereafter the disease became widespread among swine herds in Iowa and other parts of the Middle West. It persisted in various localities until January of 1919. The epizootic in the autumn and winter of 1919 was stated to be as extensive and severe as that in 1918. The disease has appeared among swine in the Middle West every year since but varies from year to year in its severity and extent.

These data, together with the one that the serum of people who were over 12 years old in 1936 served to neutralize swine flu, and that of minors (born after 1918) did not, allowed Shope to conclude (p. 684) that swine flu has human origin. : "It seems likely that the virus of swine influenza is the surviving prototype of the agent primarily responsible for the great human pandemic of 1918."

Santiago Mata: 1917 American Pandemic

The mass murderer: The second wave

Among the various mutations resulting from the 1917-1918 pandemic flu, one of them was especially deadly and spread the disease again in a particularly deadly second wave. In France, according to data from the military health service published by Françoise Bouron, the first wave, from April to the end of June, had a high level of contagion, but a brief and benign evolution, rarely fatal. Only one in eight flu cases had complications in May. In August, it would be half of the cases.

In the intervening period, references to the (first) wave of influenza in London (and Berlin) appear in the French press, and in their eagerness to attack the enemy, they even affirmed that the disease came from the Germans and did not affect the French soldiers.

On August 14, the flu appears for the first time in *The New York Times* referring to the United States, although still until October 1 the articles in that newspaper will focus on how the disease harms the Germans (or, July 27, the Swiss army). By treating it now on the cover, it would provoke a public debate on health policies, referring to a Norwegian ship that had arrived at the port of New York with ten sick passengers.

The debate on whether it was correct to break the quarantine to take the sick to a hospital lasted several days, with the newspaper accusing the authorities, in an editorial titled *Spanish Influenza* on Friday, August 16, of deceiving the public with demulcents. The unsigned comment was published on page 6 just before the first of the minor stories called Topics of the Times:

> On Tuesday a Norwegian steamer arrived at Quarantine and her doctor recorded nine cases of influenza on board. There had been three deaths from pneumonia, following influenza, at sea. Certain cases transferred from the boat to the hospital were classified as pneumonia, "probably brought on by Spanish influenza," says the physician in charge. According to Dr. Cofer, the Health Officer of the Port, the surgeon of the steamer did not use the adjective "Spanish" in reference

Santiago Mata: 1917 American Pandemic

to the cases reported on the boat. How ingenuous! At Quarantine Wednesday morning it was said that the ship had been passed because there was no quarantine on influenza at this port. On Wednesday night Quarantine said that the ship had been passed because her surgeon reported that the sick people on her had recovered.

Apparently the health officers soon became doubtful of the wisdom of letting the ship come up. Yesterday Health Department Inspectors and Port Inspector went to Quarantine to examine members of the crew and passengers who had not already passed quarantine. To some extent this was lecking the stable door after the horse had been stolen. It may be true, as Dr. COPELAND says, that "the public has no reason for alarm, since, through the protection afforded by our most efficient Quarantine Station and the constant vigilance of the city's health authorities, all the protection that sanitary science can give is assured." There may be doubts as to the protection accorded by our most efficient Quarantine Station and vigilant city health authorities if, after so many thousands of cases of Spanish influenza in Spain, in Germany, in France, in England, in Cuba, it was only yesterday that the question of laying a quarantine against Spanish influenza was taken up by the protectors. "You haven't Heard of our doughboys getting it, have you?" said Dr. COPELAND. "You bet you haven't, and you wont." The theory is that few but persons badly nourished, of low vitality, are attacked by this virulent form of influenza. But the British and French soldiers in France, some or many of whom are said to have had this influenza, are well fed. The people of England, where it has raged considerably, are well fed. The people of both Spain and Cuba

118

The mass murderer: The second wave

are enjoying remarkable prosperity, and presumable their diet is in proportion, if always soberer and more frugal than ours.

There is no necessity for alarm and nobody is going to be alarmed; but perhaps the health authorities of the port and the city have been a little too eager to reassure the public, which prefers the truth to official demulcents. And, possibly, those authorities have been too easy or incredulous in regard to our Spanish visitor.

The port's health chief, Dr. Leland E. Cofer, as reflected the same issue of the newspaper with an article on page 6 titled "Health Head calls Influenza inquiry", and subtitled "No fear of an Epidemic", stating that this type of flu was already suffered in the United States, and that in fact it would have been more appropriate to call it American than Spanish:

The New York Board of Health, at its meeting yesterday, took official notice of the fact that there is influenza, germinated in Europe, in this city. The board ordered that cultures be taken from each one of the patients now in the Norwegian Hospital in Brooklyn and that these cultures be sent to the board's bacteriological laboratory for observation and analysis. This was done as a precautionary measure and not, so Health Commissioner Copeland reiterated, because he or any member of the board believes there is the slightest danger of an influenza epidemic breaking out in New York.

Dr. Cofer, Health Officer of the Port of New York, was asked yesterday if he intended to establish at this port a quarantine against foreign-bred influenza.

Santiago Mata: 1917 American Pandemic

"I am glad you asked that question since it gives me an opportunity to clear up an evident confusion in the minds of many of the people of this city," he said. "To give you first a direct answer to your question, I do not intend to establish a quarantine against influenza, Spanish or any other kind. And right here let me digress long enough to say that Spanish influenza is altogether a misnomer. We have had epidemics of influenza in this country, with symptoms very like those of the cases developed in Spain. Yet the world didn't rise up and call it American or United States influenza. Nobody nows definitely whether this European variety that we have heard of recently is Spanish or Norwegian or Swedish or what.

"Now, you see, and you'd better make careful notation of these, there are just six diseases which are recognized the world over by health authorities, as proper diseases against which to declare a marine quarantine. They are: smallpox, leprosy, yellow fever, plague -that's the correct term, rather than a specific kind of plague- typhus fever and cholera. Just let a ship stick her nose into this harbor with a case of any one of these diseases on board and she will find herself tied up in the stiffest kind of a quarantine."

Cofer was also sincere in stating that, for the health authorities, the important thing was to supply the allies with men and war material, and that in this context the flu was one of the minor diseases that did not require quarantine:

"This country is at war and, besides having on our hands the winning of the war, we have also the job of supplying our Allies with much that they need with which to enable them to fight on successfully. This port cannot be clogged for a minute longer than necessity requires. Therefore, when

120

The mass murderer: The second wave

we find what may be called minor communicable diseases on ships entering the port, such as the ones I have spoken of, and to which must be added influenza, we simply report them to the city Board of Health, which is splendidly able and enthusiastically willing to take care of each case when or before it leaves the ship at its pier.

"There is not the slightest danger of an influenza epidemic breaking out in New York, and this port will not be quarantined against that disease."

In that same page, the official USPHS statement, dated the 15th in Washington, was published:

The United States Public Health Service is receiving full reports of the presence of Spanish influenza at New York, but will take no steps to establish a quarantine against the disease.

The next day, August 17, the newspaper reported (p. 5) that New York's Chief of Health, Copeland, had ordered the director of the department of communicable diseases, L.I. Harris, "to inaugurate, at once, a campaign against influenza and all such communicable diseases," which according to Blakely (p. 27) was "a clear case of a shift of responsibility from the New York City health commissioner to a U.S. government, or federal, problem."

Only from that moment does the obligation seem to have been established for ships to warn of cases of influenza, with a twist compared to the previous day that, for Blakely (p. 28), "demonstrates a change in framing from *there's nothing to worry about* to *we have everything under control.*"

On August 18, the newspaper reported (p. 9) that a new ship had arrived with 21 "cases of Spanish influenza". But these were not really more than what was left of the more than 200 passengers and crew who had been ill:

The liner's surgeon said that the first cases developed three days out from the port of sailing, and

121

that three days later something like 200 of the ship's company felt indisposed. The East Indians seemed to show less resistance to the disease than any of the others and five died and were buried at sea. All the rest of the sick folk were pretty ill for the usual four days and then began to mend rapidly.

The port's sanitary military chief, Colonel J.M. Kennedy, acknowledged that ships with flu patients had been arriving for two months:

Colonel J. M. Kennedy, Medical Corps, U.S.A., Chief Surgeon at the New York port of embarkation, said yesterday that some persons seemed to have notion that the city and Quarantine guardians of health had only recently taken an interest in the possibilities of an epidemic here of either foreign or domestic influenza.

"Why," the Colonel said, "the present publicity given to incoming influenza is publicity given to an old story. Ships having influenza aboard began arriving at this port six weeks or two months ago. The Healt Officer of the Port and the New York Health Department have been on the job ever since the first case was reported. They have taken all the practically possible measures that can be taken to prevent the spread of the disease in New York and to other parts of the country. I have no jurisdiction over what they do, my work being confined entirely to the port of embarkation, or the army transport service at this port, but I am quite familiar with the procedure at Quarantine and that taken by the New York Health Department.

"It has my hearty approval. Furthermore, I agree with Dr. Cofer, Health Officer of the Port, that it would be utterly impracticable to establish a

The mass murderer: The second wave

quarantine at this port against this disease. We can't stop this war on account of Spanish or any other kind of influenza. To quarantine against it would mean to isolate the patients somewhere and fumigate every ship. That would clog the harbor and produce interminable delays in the sending of troops and supplies overseas, and that cannot be permitted. 'Over there' they don't quarantine a regiment in which influenza has broken out and withdraw it from the fighting line. They take care of the sick and the rest go right on fighting. But our men could not go on fighting long if quarantines were established at American ports against this not very serious disease. Dr. Cofer is quite right in the position he has taken.

"Influenza, and the kind that is coming here in just like ours or that of any other country, is not at all dangerous, except when pneumonia develops. Pneumonia develops in a comparatively few cases, and few of those cases have fatal results, except where the patient has become debilitated through lack of proper food. The less this whole subject is agitated the better it will be for New York and the whole country and for our army overseas."

But already on August 19, *The New York Times* specified that the pandemic was not a common flu, and reported its spread throughout Russia and neighboring countries. Although there is still no quarantine, there is, according to the title of the news on page 5, an "Epidemic Guard for Port. All Incoming Steamships Watched for Signs of Spanish Influenza":

Health officers of the port are keeping a strict lookout on incoming steamships from Europe whose passengers or crews might have developed Spanish influenza. They say it would be impossible to quarantine every vessel which arrives

Santiago Mata: 1917 American Pandemic

with a few cases on board. Americans who have had both the Spanish variety and the old-fashioned "grippe" of years ago, declare that the newer form is more malignant and leaves the system much weaker, which is the cause of so many deaths from pneumonia. A considerable number of American negroes, who have gone to France on horse transports, have contracted Spanish influenza on shore and died in the French hospitals of pneumonia.

At St. Vincent's Hospital it was said that the three negroes taken there Saturday from the Holland-America liner *Nieuw Amsterdam* suffering from Spanish influenza where showing signs of improvement, but were still very ill. Some passengers from Finland said the epidemic had spread all through Russia as far north as Archangel and thence to Sweden, Finland, Norway, and Denmark.

In this, however, the travelers were wrong, since it was the soldiers of the US 85th Infantry Division who landed on September 4, 1918, who spread the deadly epidemic in Archangel. According to Willet, several soldiers mentioned having seen funerals at sea, but officially the first dead was Private Albert F. Rickert, of Mt. Clemens (Michigan), within hours of arriving at port.

The development of the second wave of pandemic influenza in Boston was summarized in February 1982 by L.S. Block in *Yankee Magazine*. The first cases occurred in the barracks called Receiving Ship, which housed 7,000 sailors awaiting orders there. Two of them arrived on the afternoon of August 27, 1918 with chills, fever, sore throat, cough, and labored breathing. Doctors on duty took blood samples and throat cultures, performed a physical exam and ordered the men to lie down. Forty-eight hours later, two of the 11 doctors in the infirmary had fallen ill.

On August 28, eight other sailors registered with the same symptoms in the infirmary; On the 29th, 58 more. By the end of the week, the number of new

The mass murderer: The second wave

cases was 150 a day, and a constant parade of ambulances took patients from the modest infirmary to the modern Chelsea Naval Hospital overlooking the Boston Harbor, with a capacity of 236 beds.

At Camp Pike, September 23 would be the official date of the second (or third, in any case the deadliest) pandemic wave, with the sudden admission of 214 cases. Since no one could believe that an epidemic arises so suddenly, Volume XII of *The Medical Department of the United States Army in the World War* (p. 130) recognizes that "the epidemic was foreshadowed by a steady increase in the number of admissions to the base hospital diagnosed as acute bronchitis. This increase began about September 1, and on September 18 there were 50 admissions with this diagnosis."

Therefore, in addition to the flu in Boston Harbor and the fact that the first camp in which the second wave was declared was - as we will immediately see - Camp Devens, Camp Pike must also be taken into account due to the extreme virulence that, again, the pandemic flu had there. The dissimulation - or the distraction - in this camp was spectacular, if we take into account that 12,393 cases of flu were diagnosed in September and October (11,899 until October 19) and another 1,499 of pneumonia, of which 466 are said to have died of pneumonia and "only two patients died with a diagnosis of uncomplicated influenza."

But the number of flu cases in Camp Pike is the same of the total number of patients, since all patients with pneumonia had flu before (12.1%) and therefore "the mortality for the epidemic as a whole was 3.8 per cent of those attacked by influenza". Even if presented as innocuous (for causing only two deaths), the described flu is undoubtedly the pandemic (p. 131):

> The influenza was characterized by sudden onset with chilliness and sharp elevation of temperature, often from 103° to 105° F [39,4-40,6° C]. there was extreme prostration, severe backache, suffusion of the face, and injection of the conjunctivae. Coryza, pharyngitis and tracheitis with a harrassing cough were almost invariable; epistaxis and slight hemoptysis, were frequent. In

Santiago Mata: 1917 American Pandemic

> the majority of cases the temperature subsided after from two to five days, usually rather abruptly. About one-third of the patients developed purulent bronchitis.

Separated by more than 400 kilometers by land from Boston - but 80 from downtown New York and less than 200 from Camp Upton on Long Island - two separate army training camps had been set up at Princeton University (New Jersey). and the navy. The first case of flu was declared on September 5 and students were prohibited from entering buildings outside the campus, disinfecting everyone who passed the campus, inspecting staff every morning, ventilating the barracks and isolating suspected cases. Of the 1,142 university students, there were 192 cases of flu (16.8%), but only one professor died. An example of how, even in its worst stages, pandemic flu could be stopped when the means were put in place to avoid the worst. In the adjacent city there were 32 deaths.

50 kilometers west of Boston was Camp Devens, with a capacity of 45,000 men (5,000 in tents), where much of the Yankee Division (26th) had trained, then the 76th (which left for France in July 1918) and since then the 12th infantry has been training. The second wave of pandemic influenza would cause 757 deaths for 13,733 registered cases of acute illness (which, in turn, meant that 5.5% of the sick died). In October the mortality rate fell to 1.04%.

At Camp Devens the flu made its comeback on September 8. Despite the risk of a pandemic, 4,000 new recruits had been enlisted in the first week of September. The hospital then had 2,000 beds, with a continuous line of soldiers joining it from the various barracks in the camp.

Army Inspector General of Health William Gorgas dispatched four inspectors to Camp Devens, who reported the sight of young men entering the hospital in groups of ten or more, with bluish faces, an agonizing cough and blood-stained sputum, and of the corpses piled up every morning around the morgue like firewood. Autopsies witnessed by these four inspectors revealed blue and swollen lungs with foamy surfaces.

One of the four inspectors was Victor C. Vaughan, dean of medicine at the

The mass murderer: The second wave

University of Michigan and director of contagious diseases. With him were William Henry Welch, a pathologist at Johns Hopkins, and Rufus Cole, an expert in respiratory diseases at the Rockefeller Institute. Among the 16 measures this team proposed were suspending troop movements to or from Camp Devens.

Before this measure was applied, a contingent of troops was sent from Camp Devens to Camp Upton, embarkation point for France on Long Island, a journey of 265 km in case they embarked for the island in New London, or 400 If they were all the way overland through New York.

On September 13, the epidemic broke out at Camp Upton, with 38 hospital admissions, followed by 86 the following day and 193 on September 15. Some patients developed pneumonia so quickly that doctors diagnosed it with the naked eye, without needing to listen to the sound of the lungs. On October 4, the maximum of 483 was reached and in total there were 6,131 admissions in 40 days (until October 22, when there were only 11 new admissions). From that date to the end of November there were another 816 cases, and the epidemic continued in an even milder manner until April 1, 1919, when three camp doctors (Major Irving P. Lyon, Captain Charles F. Tenney and Lieutenant Leopold Szerlip) wrote their report.

The mortality, according to the report, was 6.58% in the first period and 4.18% in the second (overall it was 6.3%). We can deduce that 404 patients died up to October 22 and another 34 until the end of November (438 in total).

From Camp Devens the flu spread to neighboring places: in mid-September there were 125 cases in Leicester and eight deaths in Millbury in a single day (both towns are 40 km south of the camp). On September 15, the mayor of Worcester (less than 35 km from the camp) ordered bonfires to be lit in public schools, and from the end of the month until October 7 he closed schools, theaters, cinemas and public halls (the churches were already closed by State orders), offering $ 4 a day to graduate nurses, $ 60 a month to experienced nurses, and $ 150 a month to doctors.

Santiago Mata: 1917 American Pandemic

The disease occurred in three serious forms: In the first group, it started gently and in a couple of days the patients felt better. Then there was a sudden rise in temperature, followed by the onset of pneumonia and death. In the second group, the disease began severely, pulmonary complications followed, but recovery continued. In the third group, it started out as extremely severe, breathing became very difficult and cyanosis appeared. This was the case with the 18-year-olds at Camp Devens: their lungs filled with fluid and they drowned, leaving their faces blue from lack of oxygen. Death here occurred within 36 to 48 hours.

The *Boston Globe* showed on September 11, 1918 that the civilian population in contact with the military of Camp Devens (and not with the sailors) was the first to become infected, when reporting the death of Catherine Callahan, who lived "overloaded by her work for our soldiers" and whose sister Mary, 19, was also in the hospital with pneumonia, as was her mother.

Boston's health commissioner William C. Woodward estimated 3,000 flu cases in the city on Sept. 18. 40 people had died in the last 24 hours. As of October 16, more than 3,700 people had died. Work was offered at the rate of $ 28 a week for the nurses and 15 for the assistants, plus transportation costs. In the cemetery a circus tent was placed to house the coffins that could not be buried due to lack of time.

At the end of October, religious ceremonies and festivals were allowed again, and a thousand doctors and nurses were sent to cities where the epidemic was most violent. When the quarantine was lifted at Camp Devens, 20,000 family members showed up to visit the soldiers.

By the end of the year, 4,088 of the more than 22,000 deaths from the flu in Massachusetts had occurred in Boston (the usual number of deaths from seasonal flu, 500, multiplied by eight in the city, and by 12 the usual number of 1,800 in the State). According to the US Health Service, 25 million people fell ill with the flu that winter, more than a quarter of the population of the United States. 548,000 died, according to L.S. Block.

Returning to what happened in New York, despite all the denials, for the first

The mass murderer: The second wave

time on September 13 - the same day that the epidemic was declared in the neighboring Camp Upton – an advice to the people had been published in *The New York* Times to fight against The Flu. In short, stay at home, but surrounded by a series of lies about the disease, starting with the title of the news (p. 7), *City is not in Danger from Spanish Grip*:

> Although twenty-five cases of Spanish influenza were recently taken from a ship arriving at New York persons in this city are in no danger of an epidemic, according to Dr. Royal S. Copeland, the Health Commissioner. He said that an absolutely reliable index of the possible prevalence of the disease would be many cases of pneumonia, into which the influenza develops and which makes it dangerous. As a matter of fact, at the present moment the death rate from pneumonia is half of what it was at the same time last year, and, more than that, no secondary cases have developed up to this time. That is, none of those who have arrived here with the disease have infected others.
>
> Dr. Copeland explained that when these cases were reported the patients were not allowed to land, but were taken by the department ambulances to the Willard Parker Hospital in East Sixteenth Street, where they were isolated.
>
> "None of them was very ill," continued the Commissioner, "and only two of them developed pneumonia, and these two had it when they arrived. Since they have been in our care all the patients have improved, several of them having entirely recovered and sent home. These cases have been carefully followed by the department, its physicians calling upon the patients every day. At the same time Dr. William H. Park, director of the laboratories of the Department of Health, has been studying the cases from a bacteriological

Santiago Mata: 1917 American Pandemic

> stand-point and hopes to reach some conclu-
> sions as to the prevention and future treatment
> of the disease. At present no better advice can be
> given than to tell any one who actually contracts
> Spanish influenza to go to bed immediately and to
> remain quiet. The disease will normally run its
> course in three days. So far there is no known
> cure for it."

At Camp A.A. Humphreys (westward extension of Camp –now Fort– Belvoir, Virginia, 500 km east of Camp Sherman and almost 450 in a straight line southwest of Camp Upton), doctors recorded the beginning of the second pandemic wave "around to September 13, 1918", according to the book published in 1919 by the military health inspector Rupert Blue, and lasted until September 18, with the peak of admissions on October 5 (293) and the peak of deaths on the next day (48). In total, 413 of the 4,237 admitted cases died (9.7% mortality rate among patients).

Authorities recognized the threat of a pandemic and issued orders from Military Health Inspector General Rupert Blue on September 14 for all doctors (published by D. Blakely on p. 30 of *Mass Mediated Disease*):

> Because the last pandemic of influenza occurred
> more than twenty-five years ago, physicians who
> have begun to practice medicine since 1892 have
> not had personal experience in handling a situa-
> tion now spreading through a considerable part
> of the foreign world and already appearing to
> some extent in the United States.

The cause of the warning was an "abrupt outbreak" of influenza that arose in several military training camps near Mobile (Alabama, where there were no camps), Philadelphia (Navy Yard), Boston (Camp Devens), New London (Connecticut, where there was only evidence of army enlistment barracks, which indicates that the contingent that brought the epidemic from Camp Devens to Camp Upton was able to pass through there) and New Orleans

The mass murderer: The second wave

(Louisiana), where there were several naval facilities and Camp Martin, a National Guard camp at the Fair Grounds racetrack in the heart of the city.

Flu cases were not reported at this camp until the end of the month. The *New Orleans Times-Picayune* newspaper announced them on October 1, warning that the camp was under quarantine, at the same time it spoke of four soldiers killed in Beauregard, in Pineville, almost 270 km to the northwest, a camp where the situation was "serious" and for which 40 nurses were requested.

In short, what forced these warnings to be transmitted to citizens was the epidemic that emerged on September 8 at Camp Devens and transmitted from there, for not complying with medical recommendations, to Camp Upton. In the September 14 warning, the instructions for individuals consisted of going to bed and calling a doctor, and the usefulness of quarantines was ruled out: "There is no such thing as an effective quarantine in the case of pandemic influenza." (Quote from Dr. Blue published in Volume 45 of *Municipal Journal & Public Works Engineer*, p. 20, 1918.) As Blakely notes (p. 30), the government shook off responsibility and placed it on individuals.

On September 16, *The New York Times* reported 8,000 cases of flu in Camp Devens, talking about what health officials discuss about how "to fight Spanish grip":

> Ways of instructing the public how to prevent the spread of influenza from cases that might come to the city were discussed at a conference yesterday by Health Commissioner Copeland, Dr. Louis I. Harris, Director of the Bureau of Preventable Diseases in the Health Department, and other officials of the department and the United States Public Health Service. The conference was held after the Health Department had received reports that there were 8,000 cases of influenza among the soldiers at Camp Devens, at Ayer, Mass., with about 100 cases in the naval

Santiago Mata: 1917 American Pandemic

hospital at the submarine base in New London and others in Camp Lee, Va.

Only twenty-three cases of influenza had been reported to the Health Department here, Commissioner Copeland said, and all had been isolated so that there was no danger of the disease spreading from them. All the cases were sailors from the American Navy, who contracted the illness on ships as the result of heavy colds, Dr. Copeland said, and all were mild except two that had developed pneumonia. The men were taken from ships by navy surgeons and physicians of the Public Health Service and sent to the Long Island College Hospital in Brooklyn. Four of the cases were discovered yesterday on ships from Philadelphia, Newport, Charleston, and Panama.

"There is no danger of the affection spreading here from the cases we have under treatment now," said Commissioner Copeland, "and the Health Department, the Public Health Service and the navy surgeons are co-operating to see that no new cases are introduce. A rigid examination of the crews of all ships is being made at Quarantine and the navy yard.

On September 17, *The New York Times* reported (p. 10) the order of Camp Upton commander, Colonel John S. Mallory, to quarantine the camp against the flu epidemic, prohibiting its 40,000 to 43,000 inhabitants from entering or leaving "except on the most urgent business."

In Camp Devens, the *Boston Daily Globe* reported on October 4 that Naomi Barnett, a young from Brockton (Massachusetts, 70 km southeast of the camp), came to the camp to take care of her boyfriend, Jacob Julian, upon learning that he was ill. The couple planned to marry before he left for France, but she died of pneumonia within two days of arriving at the camp, and her boyfriend died 30 minutes later.

The mass murderer: The second wave

To combat the epidemic at Camp Devens, each patient was allocated a space of 100 square feet (9.3 square meters), separating the beds with sheets and forcing all camp staff to wear a mask. To prevent pneumonia, the masks of 800 healthy people were sprayed every day with chloramine-T (a biocidal disinfectant), but after 20 days, when comparing this group with another 800 without this treatment, it was seen that the incidence of influenza in both groups was the same.

Supposedly on this day the epidemic reached Seattle, assuming a previous maritime contagion from Boston on a ship to the navy shipyard in Philadelphia, from where another ship would have taken it to the state of Washington through the Panama Canal, anchoring on September 17 in Seattle. From there, in turn, it would spread the flu in less than a month to Alaska, one of the places not reached by the first wave.

Against this maritime theory, there is the possibility that, as happened with the first wave, the transmission of the second within the United States was mainly terrestrial, and in the heat of the military camps; and even that it arose in those fields and not in Boston on board European ships, since, as we saw, the civilian population of that city was infected from Camp Devens and not by sick sailors.

In fact, the Seattle authorities denied on September 20 through the mouth of their Deputy Chief of Health, Dr. J.S. McBride, the existence of flu there, as the *Seattle-Post-Intelligencer* published the next day on its front page. Over the next two weeks, a hundred severe flu cases were reported at the Camp Lewis, 75 kilometers by road south of Seattle.

Camp doctors denied that pandemic ("Spanish") flu was spreading through the camp even after one of its soldiers was the first case of the second wave recorded in Portland on October 3. They did not declare the epidemic until the 9th (two days after it was declared in an army facility).

On September 18, an editorial in *The New York Times* had described this process as humiliating, stating for the first time that a severe quarantine could really have prevented the entry of the flu (obviously, ignoring that the second

Santiago Mata: 1917 American Pandemic

wave, like the first, could have arisen in the United States), in the first of the *Topics of the Times* (p. 12) entitled "More Likely German Than Spanish"):

> It seems now to be admitted with something like unanimity by all our medical authorities, including those of the army and navy, that this country, and all of it, is about to undergo the serious inconveniences and not inconsiderable dangers that accompany and follow an epidemic of the so-called Spanish influenza. That this new variant of the "grippe" with which we have been only too familiar in other years really originated in Spain is a matter of some doubt. It did, indeed, first come to public notice in that country, but the Spaniards declared, and the declaration is at least plausible, that the infection was brought to them by the crews of German U-boats, and they explained the peculiar malignancy of the disease as the result of the conditions in which the men in those vessels were obliged to live for weeks, and sometimes for months, at a time.

> About the best that can be said for this theory is that it is possibly true, just as it is possibly true that yellow fever was a product of the conditions in which African slaves were brought across the Atlantic to the American continents. In both cases denial and demonstration are alike doubtful.

> But whether the latest form of and old malady is Spanish or not is a question of no practical importance. We know when it came here and by what ships it arrived. We also know that the guardians of our ports at first told us that there was no danger of an epidemic here, and a little later admitted that there was not a danger, but a certainty of it, and that we would have to endure

The mass murderer: The second wave

the consequent ills as best we could, as quarantine measures, to be effective, would have to be so stringent as to work an intolerable interference with the conduct of our war businesses on land and sea.

That was a rather humiliating confession for the representatives of medical and sanitary science to make in these days, but to question its truth would not be becoming. The only advice the experts now give us is to keep our powers of resistance as high as we can by strict attention to personal hygiene. We must sleep enough, eat carefully, keep out of crowds, take lots of outdoor exercise, and abstain from fiery liquids. That is good advice. It always is and we shall probable heed it just about as much as has been our custom in the past.

On page 24 of that issue, the authorities deny the community transmission of the new flu in New York and talk about the threat posed by the flu that arrives from Boston, citing the Major Charles Ernest Goddard as camp surgeon (of Camp Dix) in the article entitled "Must Report all Spanish Influenza":

The Health Department took steps yesterday to prevent the spread here of the so-called Spanish influenza. This latest effort was embodied in the action of the Board of Health at its regular meeting in placing pneumonia and influenza on the list of diseases that must be reported by all physicians. Pneumonia occurs as the secondary stage in the disease, and is the cause of death.

Pneumonia and influenza were formerly exempted from the list, and this, the authorities held, left a way by which the disease might come into the city and its spread be attended by very little publicity until perhaps too late.

Santiago Mata: 1917 American Pandemic

Health Commissioner Copeland decided that the old method would not do in the present crisis, and he decided that the best way to check the possible spread of Spanish influenza was to put it and pneumonia on the list of diseases that must be reported.

The campaign to keep Spanish influenza from entering the city is under charge of Dr. Harris, Chief of the Bureau of Communicable Diseases. Dr. Copeland said that all precautions had been taken and that the grave danger was not that it should spread through cases arriving here on steamships, but that it might "execute a flank attack" and reach this city by the way of the railroads from Boston or elsewhere.

"In Boston," said the Commissioner, "a great many cases have been found, and naturally some of those affected will come to this city by railroad, bringing the germs with them. It would be impossible under present conditions to check up these arrivals, and the result would be an epidemic here. To prevent this we have changed the existing order."

Up to the present time, Dr. Copeland said, there have been no local cases of the new disease reported, all those found being among sailors in the service or on incoming vessels. These cases have been placed under quarantine in the city hospitals.

Three cases of influenza at 42 Clifton Place, Jersey City, were quarantined by the Health Board of that city yesterday. Dr. Joseph Craven, Superintendent of the board, said it was the old-fashioned influenza, an infectious germ disease requiring the quarantining of the patient, but, he did

136

The mass murderer: The second wave

not know of any reason for calling it the Spanish influenza or any other special name.

Major C.H. Goddard, the camp surgeon, said that many of the first cases reported last Friday were now ready to leave hospital, but that 30 new cases developed yesterday. The malady is being fought principally with sunshine and fresh air.

Every window in camp is up to stay indefinitely. The men in barracks are sleeping alternately head to foot, beds remain out doors all day, and the floors are scrubbed continually. Every man has also been provided with an extra blanket.

It was not found necessary to close the camp theatres, and the Y.M.C.A. succeeded in arranging entrance for entertainers coming from the city. The Long Island Railroad warned all persons who bought tickets in New York for Camp Upton of the quarantine, and the military police refused admission to visitors.

On September 19, the epidemic was declared just over 20 kilometers north of New York Harbor, at Camp Merritt (as stated in Volume XII of *The Medical Department of the United States Army in the World War* p. 118), registering 4,979 flu cases up to November 6, 1,015 of which led to bronchopneumonia and 31% or 6.3% of all admissions died (314 deaths).

On September 20, the first cases of the second wave of pandemic flu were reported at Camp Greene, according to the summary published in 1919 by Rupert Blue as health inspector (p. 747). The Volume XII of *The Medical Department of the United States Army in the World War* summarizes the most serious pandemic wave at Camp Greene at 4,595 flu cases, 626 of which resulted in pneumonia (but only 20 with certified pneumococcus) and 308 deaths among September 1918 and February 23, 1919.

On September 21, the second wave was already registered at Camp Grant,

Santiago Mata: 1917 American Pandemic

1,700 km from Camp Devens by road. That day there were 70 admissions. "So sudden and appalling was the visitation that it required the greatest energy and cooperation of every officer, every man, and every nurse to meet the emergency," according to one of the sources cited by Byerly in his 2010 article. Admissions rose to 194, then 370, 492 and on September 29 to the maximum of 788.

Camp Grant Hospital went from 10 occupied beds to 4,102 in six days. 11 of the 81 medical officers fell ill, and three army nurses and three collaborating civilians died. The ban on black nurses from working in the camp was lifted. For relatives who inquired about the sick, a store was set up as hospital information office.

The *Report On Influenza Epidemic At Camp Grant, Illinois, September 21st, 1918 to Nov. 5th, 1918* - Abstract From RG 112 Records of the Office of the Surgeon General (Army) [1918], five pages, one of them a graph, another with the data from the graph, and all unsigned - indicates as date for the end of the epidemic November 5, 1918 (the day before there were no new arrivals) and assures that from the beginning the barracks were placed "under strict quarantine."

The first cases on September 21 occurred "in the Infantry Central Officers Training School. The next organization affected was the 5th Limited Service Regiment, composed of September 4th draft men coming from districts in the immediate vicinity of this Camp, and where the disease had already made its appearance. This regiment was located in the same Camp Area as the Infantry Central Officers Training School."

The report claims that Camp Grant headquarters banned soldiers on September 20 from exits, except to Rockford and Milford. "The epidemic spread with a violent rapidity and inside of 48 hours most every organization in the entire camp was affected. It is the opinion of the Medical Officers is that this disease was brought into Camp Grant by the men who where transferred here from Camp Devens, where the disease already prevailed, to attend the Infantry Central Officers Training School. This school being composed of men who

The mass murderer: The second wave

came from other Camps and Cities where the men in all probability were exposed. epidemic spread with violent rapidity and within 48 hours almost all the organizations throughout the camp were affected" (that is, around September 23).

The fact that the camp had been closed in front of the neighboring towns on the 20th, and that the epidemic developed with such force in it, leads the informants to reaffirm that it was not brought by recruits from neighboring towns, but by the Soldiers from other camps:

> Further the disease broke out in this Camp before it made its appearance in nearby towns which confirms the statement that the carriers brought the infection to this camp.

On September 27, "six days after its first appearance," (page 2 of the report) the epidemic reached its climax, with 984 admissions. Visitors were only allowed to enter "on special missions or who had relatives in the Base Hospital." Leaves were suspended "except in emergency cases," which, knowing the discretion with which the military could appeal to the emergency, turned all measures against the epidemic into wet paper.

It is not easy to rate the reasons given to explain why the flu affected newly arrived recruits more than veterans:

> The organizations showing the largest number of influenza cases were the 5th Limited Service Regiment, 1st Training Unit and 13th Battallion, composed of new recruits who were more susceptible to the disease and more apt to succumb than the men who had been trained and accustomed to Army Life.

The Camp Grant report appears to have been carried out by non-medical personnel, as it does not mention that any autopsy was done, or any medication dispensed, or the symptoms of the sick, etc.

Santiago Mata: 1917 American Pandemic

THE SECOND WAVE OF THE 1918 FLU PANDEMIC IN CAMP GRANT, ILLINOIS

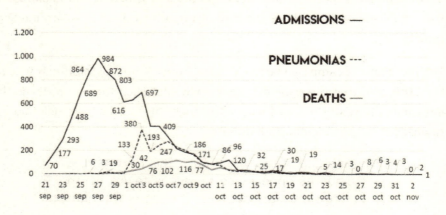

As if it were not sloppy enough that the death of more than a thousand people (1,060) did not deserve more than four badly filled pages with little data, the data sheet adds that to the 9,544 cases obtained by adding daily hospital admissions, it was necessary to add 1,185 "not reported" more, for a total of 10,729. This would mean that for a total force of 33,406 men, 32.1% were ill, that 21.8% of the patients developed pneumonia (24.5% if we did not add the 1,185 unreported cases), and 9.9 % of the patients died (11.1% without added cases). Regarding the total force, 7% would have had pneumonia and 3.2% died.

As at Camp Grant, at Camp Taylor the definitive epidemic was declared on September 21 and ended on November 15, according to the lengthy report copied in Volume XII of *The Medical Department of the United States Army in the World War* linking the population of this camp of white race with the contiguous one of black race (Camp Know) to reach a total (p. 158) of 58,000 inhabitants and confirms the paradox that the race considered superior suffered greater flu morbidity (23.1 % of patients compared to only 7% of blacks) and higher mortality compared to the population (1.6% in whites, 0.7% in blacks), although mortality among white patients was lower (6, 8%) than among blacks (10.2%), a difference for which the report finds three different reasons, none of which was that blacks received less attention than whites.

The mass murderer: The second wave

The Camp Taylor disease graph (p. 145, understood to refer only to the white soldiers camp) starts from 200 admissions in August (more than 60 were for measles) and a single death, to more than a thousand in September and more than two thousand in October, dropping below 400 in November. Measles remained at irrelevant numbers - between 40 and 100 - and the deaths were respectively 70, 150 and five. It is interesting to note that the epidemic would have a new peak in February 1919, with more than 600 admissions and more than 20 deaths.

Already on September 24, 1918, the acting chief of military health, Charles Richard, referred to the new wave of influenza, stating that "no disease which the army surgeon is likely to see in this war will tax more severely his judgment and initiative." Byerly adds that Richard's department issued flu and pneumonia bulletins for military personnel and daily reports for Peyton and other military leaders with recommendations on the epidemic, as reflected in volumes 1 and 6 of the War Department work. dedicated respectively to the role of the Inspector General of Health (*Office of the Surgeon General, Medical Department of the United States Army in the World War*, pages 998-1003) and Health in the AEF (*Sanitation in the United States and in the American Expeditionary Forces*, pages 349-371).

Richard asked Peyton not to send soldiers to France from camps where the epidemic had not ended. March approved in theory that recommendation, which initially referred to few camps. As the epidemic spread, Richard called on September 25 for all recruitment to be suspended for infected camps and for transport between camps to be minimized:

> Epidemic influenza has become a very serious menace and threatens not only to retard the military program, but to exact a heavy toll in human life, before the disease has run its course throughout the country.

March's department ordered the camp chiefs to reduce crowds and increase medical staff, but only suspended some of the recruits, so that new recruits were still arriving at the training camps at the end of September. "Only the

Santiago Mata: 1917 American Pandemic

Provost Marshall's cancellation of the October draft finally eased pressure on the camps," concludes Byerly in his 2010 article:

> Richard also recommended a one-week quarantine of all troops prior to embarkation and reducing the capacity of troopships by one-half. Desperate to build up the forces in France, March rejected these suggestions in favor of rigorous preboarding physical screening to control the epidemic. Richard countered: "It is impossible for medical officers to state with any degree of safety that any particular command is free from infection, or that it may safely embark on troopships for overseas service." He then recommended "that all troop movements overseas be suspended for the present, except such as are demanded by urgent military necessity." Richard was willing to suspend war mobilization to protect the health of the soldiers. March agreed to a 10% reduction in crowding on troopships, but that was all. The controversy reached the White House when President Wilson asked March why he refused to stop troop transport during the epidemic. March described the Army's screening precautions and invoked the exigencies of a war of attrition, pointing out (as he would write on p. 360 of his book *The Nation at War*) "...the psychological effect it would have on a weakening enemy to learn that the American divisions and replacements were no longer arriving." Troop shipments should not be halted for any reason, he told Wilson, and the president deferred to his judgment. March and Wilson had no intention of retarding U.S. participation in the war. By mid-October, however, the practice of taking men from camps that had already weathered the epidemic did finally reduce the influenza rates on troopships and in the AEF.

The mass murderer: The second wave

Sickness rates in U.S. camps ultimately ranged from 10% at Camp Lewis, Washington, to 63% at Camp Beauregard, Louisiana, averaging between 25% and 40%; death rates ranged from less than 1% at many camps to 3.3% at Camp Sherman, Ohio. (p. 138) But the sickness rates probably understated the problem because they captured only those soldiers who reported sick and received medical attention. Army investigators found that some regimental physicians did not send soldiers to hospitals unless they had temperatures higher than 101 degrees [F, 38,3° C]. (p. 3794) Many stricken soldiers may have just stayed in bed with or without knowledge or permission of their superior officers. Others may have gone home when they got sick, either with leave or AWOL. "One of the boys played wise and got sick while he was home," Charles Johnston, a soldier at Camp Funston, Kansas, wrote home in early October. "He is down with pneumonia, so will have a prolonged visit while home. Think I will try that when I come home, eh!" Several days later Johnston reported, "There have been hundreds of boys taken A.W.O.L. since [the camp was] quarantined." The situation became so bad that the War Department ordered the investigation of absentees from government service.

All these data, as it can be seen, refer to the flu within a period of time until October, and represent a moderate, if not futile, attempt to stop the second wave of the pandemic.

September 24 is also the official date of the beginning of the epidemic in Camp Sherman (Ohio), whose own news bulletin (*Camp Sherman News*) reported on October 9 that there had been 483 deaths and 1,438 seriously ill, increasing to 926 killed on October 15. Volume XII of *The Medical Department of the United States Army in the World War* provides (p. 139) the data of the population of this camp (33,044) which facilitates the statistics.

Santiago Mata: 1917 American Pandemic

According to an article published in 1966 by McCord, the total death toll from pandemic influenza at Camp Sherman was 1,052, which is 3.18 of its inhabitants. In addition, McCord places the beginning of the epidemic in that Camp in the last week of August (the 26th of that month was Monday and the 31st Saturday), but he does so without specifying sources. If true, the epidemic would not have lasted the 18 days indicated by the camp doctors, but almost a month longer, which seems implausible, since it implies that the 1918 publication concealed the August deaths:

> On a sunshiny day in the last week of August 1918, at the U.S. Army Base Hospital at Camp Sherman, Chillicothe, Ohio, 2 young, livid-faced soldiers, in the mania of delirium, rose from their adjoining beds and in feeble frenzy pummeled each other. Soon in exhaustion they fell backwards onto their respective beds. That was their last act. Both promptly died. They would have died anyway. A dozen other soldiers died that day in that ward. Where was the ward surgeon? He lay on the floor of the wardroom, unconscious from the same malady. He died next day. Where was the ward nurse whose duty it was to protect all her charges? She lay in delirium on a bed in the nurses' quarters. Where were the ward orderlies? They, in their duties, were attempting to rustle just a few items of hospital linens and such, so that dying soldiers at least might die on something better than a bloodsmeared army mattress.

The official study of the epidemic at Camp Sherman was published in the Nov. 16, 1918, issue of the *Journal of the American Medical Association* by Major Alfred Friedlander (of Cincinnati, chief of the camp medical service) and three other physicians (Elder Carey P. McCord and Captain Frank J. Sladen, both from Detroit, and Lieutenant George W. Wheeler, from New York).

The mass murderer: The second wave

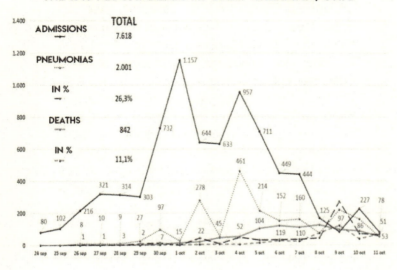

It seems clear that the pandemic did not begin or end in the 18 days of this study (the first day there were 80 admissions, and the last still 78 new ones). The Camp bulletin dated four days later increased the fatalities by 84, which is not surprising, in view of the fact that the day the study was suspended there were still 53 dead. In those 18 days, the pandemic had a daily average of 423 admissions, 111 pneumonia cases and 47 deaths.

The 7,618 flu cases (up to October 11) mean that 23% of the staff were seriously ill (required hospitalization) and that 11.1% of the sick (3.2% of the inhabitants) died. In addition, there were 3,361 cases of influenza not admitted to the hospital, so the incidence in the population exceeds 33%. Other sources, as we saw when talking about Camp Custer, give a mortality rate of 13.14%.

The article details the severe form of influenza that constituted "the outstanding clinical feature of the epidemic":

> This formed a distinct clinical picture not emphasized in any published reports. During the height

of the epidemic, many patients exhibited on admission a strikingly intense cyanosis, especially noticeable in the lips. This was not the dusky pallid litheness to which one is accustomed in a failing pneumonia, but rather the deep blueness characteristic of methemoglobinemia. These patients had high fever, intense air hunger, complete exhaustion and prostration. They were semicomatose or in a low, muttering delirium. The lungs contained diffuse bubbling rales, increasing rapidly in number and extent, in addition to subcrepitant rales. The course was rapid to death in twenty-four or forty-eight hours. The patient was practically a drowning man. The picture resembled an acutely progressive pulmonary edema. With the increasing moisture in the lungs, however, there was no sign of myocardial insufficiency or dilatation. The pulse was fair in volume and tension. Cardiac outlines were unchanged. There were neither enlargement of the liver, nor serous effusions or edema in other portions of the body.

These clinical observations were supported by the necropsy findings in these cases.

At necropsy, those dead of the condition designated clinically as an acute inflammatory pulmonary edema presented lungs having one or more lobes dark red or bluish gray, firm and rounded, with no tendency to collapse. The pleural surface was smooth and glistening, not thickened, without exudate. The lung tissue pitted deeply on pressure. The process was essentially massive and confluent. There was no evidence of a lobular distribution. Section through an involved lobe revealed an extreme grade of congestion and edema. Immediately on section there was a free outlow of thin, dark red fluid from the cut surface.

The mass murderer: The second wave

From 150 to 200 c.c. of this fluid were treasured from a single section across the lung in the different cases. The cut surface was somewhat rough but not granular; there was no evidence of fibrinous exudation. Stained films of this thin fluid showed large numbers of red cells, very few leukocytes and epithelial cells, and many gram-positive cocci in pairs and short chains. The appearance of the bronchi was the same as that described in the general section of pathology below. The pericardium was normal throughout. The pericardial cavity contained from 25 to 30 c.c. of clear, straw colored fluid. No portion of the heart evidenced any enlargement. The myocardium presented normal color and consistency.

Camp Sherman doctors insisted in their report on the surprising fact that more than two-thirds of those hospitalized (69%) were soldiers who had been in the camp for less than a month, representing less than half the population. The influenza bacillus (*bacillus influenzae, Pfeiffer*) "has not been proven as the causative organism. The frequency of its detection has not exceeded the frequency of its existence in normal conditions ", on the other hand, pneumococcus," mainly type IV "was the" predominant organism "and appeared in 53% of autopsies.

Regarding the treatment of the sick, the doctors refer that they kept them in the hospital until seven days after the peak of the disease, and the measures taken in the camp by the committee to which the commander gave "full powers" and of which two of the doctors were members, were: quarantine, prohibiting the soldiers from leaving, "and letting only the relatives of very sick patients enter", closing common places and prohibiting meetings of soldiers in closed places, cleaning barracks and avoiding overcrowding of healthy soldiers, ventilation, aeration of clothes and sheets, execution of gargles twice a day with a liquid against pneumococcus; street sweeping twice a day to remove dust. The text ends without conclusions on the effectiveness of these measures.

Santiago Mata: 1917 American Pandemic

Also in Camp Cody the epidemic began on September 24, according to Volume XII of *The Medical Department of the United States Army in the World War* (p. 31) when a soldier from Camp Dix (New Jersey) was admitted to the hospital, and said that he had been ill for three days. The soldier would die on October 1. In this camp, 75 of the 100 nurses who helped fight the epidemic fell ill, and four of them died.

The doctors' requests to limit the transport did not take effect due to the urgency of sending soldiers to Europe, and so on September 25, 3,108 soldiers embarked by train from Camp Grant to Camp Hancock (Georgia), a journey of more than 1,500 kilometers . Upon arrival, two thousand had to be hospitalized with the flu.

In fact, it appears that two thousand soldiers were sent to Camp Hancock and as many to Camp MacArthur, according to Benjamin S. Deboice's testimony presented by Opdycke (p. 170-171):

> The flu struck Camp Grant along in the early fall of 1918. We knew what we were up against because it had hit the Great Lakes Naval Training Station in Waukegan, Illinois, before it hit Camp Grant. And they died like flies at Great Lakes... It was early in the fall of 1918 as I said, just at the time when they had had six deaths at Camp Grant. I was ordered to take a troop train to Camp Hanckock, Georgia, [part of] a troop movement that involved two thousand troops to Camp Hancock, Georgia, and two thousand troops to Camp McArthur, Texas. I had five hundred and some odd men on my train.
>
> When i went up to the headquarters to get my orders, travel orders, I met for the first time the... captain of the medical corps who was to accompany our train, and i told him "You know if this train moves we're going to have a flu epidemic on that train." And he kind of pooh-poohed the idea

The mass murderer: The second wave

that touched me off and I said "Well, if you're going to treat it that lightly I'm going to give you my first order right now"...

I was a lieutenant and he was a captain, but the line officer commands. Now I was in command of the train. I said, "[When] you report for duty on this train, I want you to come with three times as much compound cathartics [*a laxative*] and aspirin as you think we can use for five hundred men on this train this trip." When he showed up at the train he showed me a pint bottle about three-fourths full of compound cathartics... and a box about three inches square and six inches tall half full of aspirin tablets and says, "Well I followed your orders." Kind of made light of it.

Well he retired to his state room and we were loading the troops and my troops were examined. They'd been examined three times in the last twenty-four hours, and as they filed onto the train the doctors watched them and anybody with a flushed face was pulled out of the line. They pulled six men out of the line with a flushed face and put a thermometer in their mouth, and they had a temperature and they fired them back so I lost six men before I started... But as we progressed towards Chicago on the train they began showing up and I... found myself walking up and down the aisles and spotting fellows that were beginning to show fever and isolating them in the back coach. I followed that procedure until we got to Evansville, Indiana, along about midnight that night. I was just completely worn out, I had a soldier trailing me with a bucket of water and I was feeding them compound cathartics, aspirins, and water.

Santiago Mata: 1917 American Pandemic

Well, I went to bed around midnight and the next day it continued again, but they got to the point where I couldn't isolate them. There were just too many on the train. We got down to Memphis -and incidentally I was carrying the bottle of aspirin or the bottle of compound cathartics and the box of aspirin. My doctor stayed in his state room. I had a very low opinion of him, but I won't go into that...

We got down to ... Chattanooga and I telegraphed ahead to the Red Cross for another supply of compound cathartics and aspirin... And I telegraphed ahead to Camp Hancock that I'd be in the next morning at seven o'clock with two hundred cases of flu on board, and I had somewhere between five hundred and six hundred men on my train. Well, no commanding general ever got the reception that we got when I showed up there... When we got there the commanding general was down at the train to meet us and all of the ambulances that they had in the camp. [...]

I started up the train to the car where my state room was... and a runner caught up with me and said, "General so and so wants to see you."... When I reported he said, "Are you in command of that train?" I said "yes sir." "Well what the blankety blankety, blankety, blank, blank, blankety, blank did you bring that outfit down here for?" Well it was so utterly ridiculous asking a second lieutenant why he moved troops around that I was just thoroughly disgusted. I moved up close and breathed all over him and I hope the got the flu.

As for the train bound for Camp MacArthur, it would spread the epidemic not only to that camp, but also to some locality along the way, as the same testimony just quoted reveals:

The mass murderer: The second wave

Another train load that was heading from Camp MacArthur, Texas, came through my home town of Clinton [Illinois]. It happened on that particular day that they were having a liberty bond program to boost the sale of liberty bonds and a general town celebration. A cousin of mine, Harry McDonald, was sort of in charge of the program and he heard that this troop train was coming through Clinton, and he thought I might be on it, so he telephoned down to the depot and told them to get ahold of the commander of that troop train and invite him to parade his troops in the parade in Clinton. Well that train was just like the one I had. It had flu, I mean it had flu on it.

On September 26, the epidemic was declared at Camp MacArthur (Waco, Texas), where the second pandemic wave lasted 33 days, according to Dr. Leon Medalia in the article he published in March 1919. Patients who did not have the bacillus of Pfeiffer were rejected from the hospital (to the extent that their conditions allowed), so one can get an idea of how ineffective containing the epidemic was. In total, 2,279 soldiers fell ill, of whom 1,752 were admitted to the hospital (77%, that is, those who were found to have Pfeiffer bacilli in sputum; therefore, almost a quarter of probable flu patients did not enter) and 61 autopsies were carried out (if this figure is equivalent to the number of deaths, 3.5% of those admitted and 2.7% of the patients died) as of October 3, when the first death occurred.

Back in the civilian world, on September 28, the 27 deaths recorded in New York in just 24 hours equaled the total accumulated since July 1, and since the deadliest moment of the pandemic began, American newspapers would dedicate themselves in forward to present statistics of deaths and affected more than other considerations.

On September 29, the start of the second wave is officially recorded at Camp Dodge (where the first wave has already claimed a hundred soldiers' lives). By the time the quarantine was imposed there, 1,500 were sick and there was one dead (according to the October 4 issue of the camp weekly). In all,

Santiago Mata: 1917 American Pandemic

13,700 soldiers were admitted and 700 died before the armistice ended the war.

Volume XII of *The Medical Department of the United States Army in the World War* (p. 72) specifies that the diagnosed cases were 573 in September and 6,480 in October. Unlike the places where it is specified from where the new wave arrived imported, here it is supposed to have appeared suddenly in three different points of the Camp Dodge hospital:

> During the week preceding the sudden appearance of the epidemic of September 28, 1918, three distinct outbreaks of an infectious nature occurred in widely separated sections of the base hospital at Camp Dodge. Pharyngeal cultures from these cases showed *Streptococus hemolyticus* and *Bacillus influenzae* presents in unusually large numbers. Admissions to hospital during this period increased moderately in number, and an unmistakable but not alarming number of acute nasorespiratory disturbances, not unlike similar clinical conditions of the preceding month, gave warning of impending trouble.

On the same day, September 29, the first case of "Spanish flu" was declared in Camp Custer and the subsequent quarantine began, as reported on October 3 by the newspaper Trench and Camp. The epidemic would affect 25% of its inhabitants and with 533 deaths, which represented 5.95% of the sick (then these were 8,958), well below the death rates of Camp Grant (8.66%), Camp Devens (9.78%) and Camp Sherman (13.14%). In the state of Michigan, deaths from the flu were 6,742 in 1918, followed by another 3,000 in 1919 and as many in 1920.

Volume XII of *The Medical Department of the United States Army in the World War*, edited by Charles Lynch, says that the death rate from the epidemic in September was 5.38% of the sick at Camp Custer and in October 8.34%. The summary of the study published by Blanton and Irons on Camp Custer in December 1918, avoiding mentioning the death toll, lavished on

152

The mass murderer: The second wave

laboratory analyzes that would prove that *streptococci* causing pneumonia were present in all deaths, although it has the honesty to clarify that those were not the cause of the disease (p. 12):

> The patient is attacked by a severe acute infection of unknown etiology. In the prepneumonia stage the principal objective signs are those of a marked mucopurulent bronchitis, with high temperature, generalized pains, prostration and a leukopenia. If the process advances, the peribronchial lung tissue is quickly involved by direct extension. A massive bronchopneumonia, in may cases accompanied by acute pulmonary congestion and edema, rapidly develops, which as quickly resolves or goes on to an early termination. The marked leukopenias are probably only one phase of a general overpowering of the defensive forces of the organism. The rarity of concurrent septicemias is doubtless explained by the rapid course of the disease and the mode of origin from the bronchi. The influenza bacillus has played a minor rôle in the production of bronchopneumonia. The infective cause of the antecedent respiratory infection remains undetermined.

The 85th Infantry Division, which had been in formation since Camp Custer opened (it was activated on August 25, 1917), received in June 1918 a final replacement of recruits from the Midwest (mostly from Michigan and particularly from Detroit, including many Eastern European immigrants or children of immigrants). This division left Camp Custer on July 14 and arrived at Camp Mills on Long Island, New York on the 15th. Two days later, President Wilson authorized sending a force to fight the Soviets in Russia.

On July 21, the men of the 85th boarded the *SS Plattsburg* (US) and *HMT Northumberland* (British) transports, which set sail the next day in a convoy of 11 to 18 ships, which landed in Liverpool on 3 and August 4th. On August 27, they boarded the *Nagoya*, *Somali* and *Tydeus* transport ships, where the

Santiago Mata: 1917 American Pandemic

flu epidemic was allegedly declared. This happened particularly with the Italians who boarded the *Tsar*, a ship that moored in Murmansk. The Americans did it on September 5 in Archangel: the *Somali* with more than a hundred cases of flu, the *Nagoya* with 75, the *Tydeus* with none.

Protective sequestration, the West and Overseas

In contrast to the multitude of negligence, in some places it was possible to prevent the spread of the epidemic, resorting to measures that a study by Howard Markel and three other researchers from the University of Michigan School of Medicine called "protective sequestration" (p. 5), and that at least was applied in one of the military camps.

The seven locations included in this study were: Yerba Buena Island (San Francisco Training Station); Gunnison (Colorado); Princeton University (New Jersey); the Western PA Institution for the Blind, a college for blind boys in Pittsburgh (Pennsylvania, with 179 students, of whom 12 had the flu and none died); the Trudeau tuberculosis sanatorium in Saranac Lake (New York, with 356 patients and no cases of influenza); Bryn Mawr College (a girls' college in Pennsylvania, with 465 residents, of whom 100 had the flu after September 26, but none died); Fletcher (a town of 737 in Vermont, with only two flu cases); and Camp Crane (Allentown, Pennsylvania).

The study lists the measures, consisting of quarantining any outsider who tried to enter, self-sufficiency of vital supplies, conviction and imprisonment of offenders, and the ability to maintain a normal life, all of which required the presence of leaders above the ordinary.

According to the study, such measures were impossible to comply with in large cities, which is why "protective sequestration" was sometimes chosen in areas or educational or health institutions within an urban area. In the future - which was the purpose for which the study was commissioned - one might think of subjecting key institutions for education and national security to such kidnapping, but in 1918, obviously, it was the army that most dramatically failed in such protection. As a drawback, it is pointed out that populations that are saved from one pandemic wave are more vulnerable to the next.

In any case, protective sequestration was the only nonpharmaceutical intervention (NPI) that served to save some population groups from the pandemic. Five other NPIs were insufficient: 1) to isolate sick people; 2) quarantine sick people and those who had contact with them; 3) close schools and other meeting places; 4) reduce the risk of infection with face masks, hand

Santiago Mata: 1917 American Pandemic

washing, etc; and 5) public health information campaigns.

Particularly interesting is Camp Crane, which had been established as a Health Service camp, whose preservation of the pandemic proves to some extent that, had it been able to act freely, the health service would have had a better chance of limiting the epidemic. Although that camp had reached 10,000 inhabitants, on average during the second flu wave it had 2,119. The first of the 355 cases of influenza that it registered was on September 26 and there were 13 deaths, which means that 16.7% of its inhabitants were sick and 0.6% of them died (3.7% of the sick).

At Camp Crane beds were spaced, all public utensils were boiled, beds were installed awnings to control airborne spread, and the first cases of fever symptoms were promptly isolated. Shelves and temporary walls were also installed between the beds to break the air flow. In addition, the camp command organized the drainage of the sites, the installation of large centralized hot water boilers for showers and cleaning inside most of the buildings and secondary boilers for the dining room.

In the naval base of Yerba Buena (6,000 inhabitants) there were no cases of influenza before the "protective sequestration" was lifted: after December 6, 1918 there were 25 cases of influenza and five deaths (mortality of 20% of the sick , which illustrates what has been said about the vulnerability of an overprotected population).

Meanwhile, the second wave of the pandemic flu had reached South Africa with a contingent of 1,300 native workers repatriated from France, aboard the ships *Jaroslav* (which arrived in Cape Town on September 13) and *Veronej* (which arrived on September 18). According to Killingray and Phillips, "Mortality rates in Africa were higher than in Europe, varying between 2 and 5% of the population. In South Africa, the official death toll of 140,000 markedly underestimates mortality among Africans," modern studies suggest that the figure may be double. In Kenya, up to 150,000 people died, representing 5.5% of the population.

Also in September the second wave made its appearance at Camp Funston

Protective sequestration, the West and Overseas

and the whole of Kansas. As Lori Goodson wrote on March 1, 1998 in *The Manhattan Mercury*, that month there were 133 flu cases throughout the state, going to 1,100 in one week and by mid-October to 12,000. In the first three weeks of October, 14,000 military personnel fell ill at Camp Funston and 861 died. By the end of the year, the death toll in the state of Kansas was 12,000.

In Spain, as published by *El Heraldo de Madrid* on September 30, 1918, the Provincial Board of Health of the capital indicated as the main victims of the flu "weak people", namely: "those with tuberculosis first of all, colds, heart patients, diabetics and all those who suffer from a chronic process of noble heart are its victims, especially among those who, in addition to being chronically ill, are elderly ", confirming that it was spread" by direct contagion, especially living in confined atmospheres where there are sick people ", but not because of the water, and it was stated, with common error, that" it is due to the Pfeifer bacillus ".

The board noted the possibility of transmission to or from animals, without specifying which species, and that "it is accompanied by very varied pains throughout the body and intense prostration that persists for a long time, even in relatively mild cases. The proportion of those attacked in cases of epidemic does not drop from 40 to 50%." By classifying his "forms" he contradicted the aforementioned that he attacked the weak:

> Cerebral: congestions, meningitis, paralysis and even madness, etc. Thoracic: pulmonary congestions and pleurisy, bronchitis, etc. Abdominal: gastric catarrh, typhus infections, appendicitis, jaundice, etc. Apart from various neuralgial and other extremely varied locations. There are vulgar forms that can lead to syncope. The generalized form of the flu generally attacks robust organisms. It begins with headaches and pain in different parts of the body, lumbago, chills, tingling, itching, nausea, bilious vomiting, sometimes copious sweats and decreased urine.

Santiago Mata: 1917 American Pandemic

The board asserted with a new contradiction that "the systematic isolation of the sick is not an effective measure to prevent the disease, even though it is convenient that only the people who care for them contact them" and that the prophylaxis consisted of "treating with more care than usual for asepsis and cleaning of the oral cavity and nasal passages (...). Recommend walks in the fresh air and living in pure atmospheres that are not confined. Avoid contact with the sick and their belongings. Disinfect their clothing and bed, prudently renew the air in the rooms in which they are.

Regarding medication, it was recognized that "there is no true specific for influenza, even though quinine provides, in most cases, recommended services, mainly as a tonic, and soda salicylate, aspirin and many others".

On October 4, 1918, as *El Liberal* published the following day, the Inspector General of Health, Mr. Martín Salazar, confirmed that the epidemic that was suffered was flu, "with more virulent characters than last spring" although "it does three days that a decrease in the flu is being noticed", that it was advisable to "avoid agglomerations" as a first prophylactic measure, in addition to" being prevented by washing the nose, mouth and larynx with aseptic substances "; that "rigorous precautions were taken at the border to avoid contagion" and that "this disease, which has traveled all over the countries of Europe, came from New York", implying that its spread was not due to Spain being less developed than other countries:

> In Sweden and Switzerland, despite their great media and health culture, flu has not been prevented. In Switzerland, it also attacked, as in Spain, the Army in large proportions, and public opinion must reassure itself knowing that within the limitation of being a diffusing element such as air, the health authorities do everything they can to dominate the epidemic.

Contrary to what was done on the public fronts, according to Porras (p. 659), Spanish doctors did attribute the spread of the disease to the insufficient sanitary level of the country, but these were unjustified claims:

Protective sequestration, the West and Overseas

This attitude was an expression, on the one hand, of the desire of the doctors to combat the criticism they were receiving and that blamed them, even in part, for the magnitude of the epidemic, and, on the other, of the desire of this group to take advantage of the conjuncture to vindicate some of their aspirations for improvement at the professional level. Given that Madrid doctors, a year before the 1918-19 flu epidemic began, considered that their profession had been discredited.

For their part, the public proclamations were late in the first wave and non-existent in the second in Madrid (p. 661-662):

During the first outbreak, the municipal and provincial authorities accepted that there was an epidemic in Madrid when it had reached great importance and the press had already reported on its existence. However, during the second outbreak, the true situation was concealed, especially by the municipal authorities and only when the peak of the outbreak was recorded was its presence or, more precisely, the existence of some cases admitted. (...) While in the first outbreak the governmental and legislative authorities had no problem recognizing the existence of an epidemic in Madrid, during the second they offered resistance to admit its presence and accept that it was the flu. The Minister of the Interior transferred responsibility for the epidemic to Medicine and, especially, to doctors.

The lack of transparency - or simply ignorance - demonstrated by government officials, doctors and the media, which according to Porras always gave priority to not having social alarm, was punished by the public (p. 666-667):

The population criticized the actions and attitudes of the doctors and, more harshly, those of

Santiago Mata: 1917 American Pandemic

the political and health authorities during the epidemic. Since most of those affected were young adults, working life came to a halt at the peak of each outbreak, especially during the first.

Meanwhile, the second wave of the pandemic in India had also made its appearance in September. From Bombay and transported, like the first wave, by soldiers returning from the front, it would reach northern India and Sri Lanka in October.

Assuming that the flu will cause between 20 and 50 million deaths worldwide, Chandra and Kassens-Noor accept the figures of between 10 and 20 million deaths from influenza in India, "and a point estimate of population loss 13.8 million for the British-controlled provinces", while Killingray and Phillips accept a figure of between 17 and 18 million deaths in the subcontinent.

Back in the United States, the second wave of the pandemic reached the army training camps on the West Coast in October 1918 as well, although its presence was first recognized in a civilian institution. It was on October 3 at the hospital in Portland, the capital of Oregon, where arrived Corporal James McNeese, who from Camp Lewis (200 kilometers north by road, in Washington) was heading to his destination in the officer training camp of cavalry in Leon Springs (Texas), distant almost 3,500 km from his camp of origin.

McNeese was sent to the Vancouver Barracks military hospital in Washington state, with the diagnosis of Spanish flu, but in his camp of origin the epidemic was not declared until October 9, a day later than in Camp Fremont (California).

In Seattle, on October 4, the press reported that the day before (at the time McNeese fell ill, although they did not know it) a cadet from the naval training station established at the University of Washington died and 700 others were sick of influenza, 400 of them being hospitalized.

That same afternoon, two cases of flu were declared in the city and the state commissioner of health T. D. Tuttle ended the fiction of differentiating it from

Protective sequestration, the West and Overseas

the pandemic classified as Spanish. Mayor Hanson pointed out that the only way to avoid the fate of the eastern cities was for strict quarantine measures to be applied individually, as it would be futile to enforce it, but at least the shows were banned and the vehicles were ordered to be ventilated in public transport.

In mid-October there were more than 3,500 cases in Seattle and on the 29th, the mandatory use of face masks came into force. Until February 1919, deaths from pandemic flu were 1,400.

Despite what happened in Seattle, it was already seen that Camp Lewis would be highlighted as the camp with the lowest morbidity (10%) in the deadliest wave of pandemic flu. The data seems manipulated, since according to Volume XII of *The Medical Department of the United States Army in the World War* (p. 98) 7,088 cases of influenza were registered there plus another 1,126 of bronchopneumonia (among those of influenza 2,730 would evolve to bronchitis and 858 to bronchopneumonia).

If we take into account the desire to discount cases of influenza - in this case supported by the fact that the doctors only found Bacillus influenzae in 243 of the 1,406 samples studied- and we assign to it the 1,126 cases of bronchopneumonia, we have a total of 8,214 which, giving a 10% morbidity, would require a population of 82,000 that Camp Lewis did not have. If the 45,000 inhabitants that it had in November 1917 were maintained, the morbidity would have exceeded 18%. With a total of 152 deaths, the mortality rate was 1.8% of the total of 8,214 cases.

Camp Lewis thus showed a lower mortality than other camps, assuming 2.25% of cases in September 1918 and 3% in October, but falling, when its increase was most feared, to 0.7% in November , month in which physicians found hemolytic streptococcus in only one sample (and never again in the entire epidemic).

More than a thousand kilometers to the south, at Camp Fremont (California), the second pandemVolume XII of *The Medical Department of the United States Army in the World War*ic wave was described in epic form, according

Santiago Mata: 1917 American Pandemic

to the account in (p. 75):

> The expected storm of the prevailing pandemic infection broke suddenly on the 8th Division, at this camp, October 8, 1918, and during the next 6 weeks 2,418 patients suffering from respiratory diseases were admitted to the base hospital. In addition, many soldiers having more or less mild infections were cared for in the various camp infirmaries in order to avoid overcrowding in the base hospital. Altogether there were, at a conservative estimate, 3,000 cases.
>
> Pneumonia was diagnosed in 408 cases, an incidence of nearly 14 per cent. We know now, however, that there were many cases of pneumonia that were diagnosed as bronchitis, and that the true incidence of pneumonia was greater than that indicated.
>
> Of the 408 patients with pneumonia diagnosed, 147 died, a mortality of 36 per cent for the pneumonia series and about 5 per cent for the epidemic. No deaths occurred without a complication of either lobar pneumonia or bronchopneumonia.

October registered, according to Byerly, the highest mortality in US military installations: on the 4th in the United States and on the 11th in the expeditionary force in Europe, the incidence of the disease being more than double in the camps than in the AEF, because in the former 36.1% of their men fell ill and in the AEF 16.7%.

The difference, supposedly, would be explained because the AEF personnel would have been exposed to the first wave, acquiring some immunity, while the novice recruits from the camps were not. In the AEF, the soldiers hospitalized for war wounds were 224,000, while 340,000 were for influenza.

Meanwhile, the second wave of flu was reaching New Zealand, a country that

Protective sequestration, the West and Overseas

had not suffered the first. It is generally assumed that the epidemic arrived aboard the RML *Niagara* on October 12, 1918. The *Niagara* was just one of dozens of ships arriving with troops back from Europe. She had sailed from Vancouver and San Francisco and on board were New Zealand Prime Minister William Massey and his Vice President Joseph Ward, returning from Great Britain. Massey insisted that he and Ward be treated like everyone else regarding quarantine measures.

29 crew members and several passengers from the ship were hospitalized in Auckland, but doctors said those cases were no more serious than others already seen in the city, where six people had died in the previous days. The explosive epidemic, which used to emerge 48 hours after infection, took even two weeks to appear. As of December, there were 8,600 dead, including 2,160 Maori. Military camps were the only places punished with uniform severity.

The second wave of the pandemic appeared in Alaska - where it was also actually the first wave - on October 14 in the capital, Juneau, in the extreme southeast of the territory, and on October 20, with the arrival of the ship SS *Victoria* from Seattle, in Nome, south of the Seward Peninsula and more than 3,200 km by sea from the capital. Facing Nome, south of the Gulf of Norton and on the banks of the Yukon, is Mountain Village, one of the towns that would get rid of the flu. Another one, Shishmaref, is on a small island off the coast north of Nome, at the opposite end of the Seward Peninsula. Between the two and only 100 kilometers south in a straight line (and just over 200 by sea) from Shishmaref, was the Teller Mission (today called Brevig).

In fact, Nome was mired in a crisis that had just deprived her of the hospital founded by the Sisters of Providence, a congregation of Catholic nuns founded in Indiana in 1840, who established a hospital in 1902, when Nome added 20,000 miners who had come in looking for gold. With the mining industry gone, the nuns had to close their hospital on September 20, 1918.

Historian Laurel Downing Bill summed up the 1918 epidemic in Alaska's *Senior Voice* on April 1, 2015, stating that it came "despite the fact that the territory's governor, Thomas W. Riggs, did everything he could to keep it away

Santiago Mata: 1917 American Pandemic

from the shores of Alaska." So, learning that 75 people had died of the flu in Seattle in the week of October 12, Riggs asked the shipping companies to screen all passengers who boarded the last boats of the season and not allow those with symptoms of flu to travel. If someone arrived with them, they would be isolated at their landing port by the medical inspectors.

By the end of October, there were 350 dead in Seattle, but it didn't take that long for the first case to be declared, on June 14th. The authorities asked people not to congregate in churches, schools, social gatherings, etc. For the Seward Peninsula, as we have seen, the pandemic arrived aboard the last ship of the season, the SS *Victoria*, anchored in Nome, despite the fact that three doctors separately inspected all the crew and passengers before leaving Seattle.

Upon disembarking, none showed signs of the flu, despite which all were subjected to five days of quarantine at the Santa Cruz hospital (recently abandoned by the Sisters of Providence). Cargo and mail were fumigated.

After five days, one of the newcomers in the SS *Victoria* fell ill, but the doctors diagnosed him with an attack of tonsillitis and suspended the quarantine. Four days later, a hospital employee died. Two days later, the local chief quarantined Nome, prohibiting its inhabitants from leaving the municipality. During the return trip, on November 25, 31 people died aboard the SS *Victoria*.

Unknown to Nome, the natives began to fall ill and die. In 80 days, 162 died. The people of Mountain Village made the decision to isolate themselves. "For the natives, it was very hard," explained the school teacher, "but the safety of their own families led them to break their habits and the customs of their ancestors (of spending time together)."

Within two weeks, the superintendent for Eskimo education, Walter Shields, had died. He was replaced by a 37-year-old teacher, Ebenezer Evans, who reported that "walking through its streets, Nome looked like a dead city: half of its white population fell ill and half of the natives died. Panic struck the natives, whose fever suggested they should seek cold air. They would get out

Protective sequestration, the West and Overseas

of bed sick and go for cold air, which caused pneumonia that took them away quickly. Every day an average of 10 to 20 natives died in Nome, and the hearse was constantly busy. Many froze at night when the fire went out."

Rigg's orders were not innocuous, since according to Bill some of the couriers who carried his message also transmitted the flu to York (58 km west of Brevig) and Wales (another 22 km west, at the end of Seward peninsula), where 170 and 310 people died respectively. The natives who helped with their dogs to unload supplies in Teller, when returning to the mission in which they lived 10 km to the north, carried the virus, and from November 15 to 20, 1918, 72 of the 80 inhabitants died. With that 90% mortality rate of its inhabitants, that town that today is Mision Brevig would be the ideal place, many decades later, to recover the H1N1 virus from one of the corpses kept in permanent freezing in the common graves excavated in permafrost.

On Kodiak Island, south of the western end of Alaska Bay, 47 of its 550 inhabitants died. The last populations to be affected, in the spring of 1919, were Fairbanks (in central Alaska, 845 km east of Nome), the Aleutian Islands, the archipelago that enters the Bering Sea from the south of the bay from Bristol, and this very bay. That summer, the disease disappeared, having killed 150 whites and 1,500 to 2,000 natives.

Miners sent by Evans to inspect towns on the Seward Peninsula, with temperatures below minus 45 degrees Celsius, returned saying they had seen frozen corpses huddled together and dogs fighting over human limbs, dazed children searching for their families, and in a town north of Nome, a frozen man hugging a stove. A team arrived in Shishmaref before the flu, and its inhabitants posted armed guards 13 km from the town, to prevent anyone from entering. In contrast, in the mission north of Teller, there were hardly any survivors left.

In Wales, the disease was imported by a postman who arrived in early November with his sick son. Three weeks later the envoys from Nome arrived, finding children still suckling from their dead mothers and a girl who shivered as she warmed containers of milk between her legs to feed her siblings. The 120 survivors - including 40 orphans - were gathered at school to feed them

Santiago Mata: 1917 American Pandemic

reindeer soup. Wales ceased to be one of the main native villages: with no time for funerals, the rescue team dug two graves with dynamite and buried 172 bodies plus an unknown number of members and killed 45 starving dogs. According to the article on Wales published by Tony Hopfinger in the *Anchorage Daily News* on May 27, 2012, Superintendent Evans documented this in a report dated 1919:

> On entering Native igloos, in some cases, bodies were found in an advanced state of decomposition, where the adults had died and the children or women had attempted to keep the fires going. In many cases were found living children between their dead parents, huddling close to the bodies for warmth; and it was found in Wales that live dogs, taken into the house for comfort, had managed to reach the bodies of the Natives and had eaten them, only a mass of bones and blood evidence of their having been people.

Evans proposed that the orphans from Wales be adopted from other towns. A year and a half later, a Presbyterian missionary named Henry Greist wrote what Evans had told him: that when visiting the town, he presented the survivors with the option of taking the orphans or marrying the widowed men to the widowed women and rebuilding the families by distributing the orphans. Without further discussion, he put men and women on different sides and asked the men to choose a woman, after which his secretary married them: when a man doubted, the secretary chose him a wife. Then the marriages were formalized and the children distributed, after which, Greist concluded, "misfortune hovered over the town for years."

The greater mortality of the natives was not only due to their immune weakness, but because, when they were sick, they could not keep the fire and entire families froze, or they could not look for food and they died of hunger: some came to eat their dogs, something that seemed unthinkable, and in other places the opposite happened. In Haines, just over a hundred kilometers north of Juneau, 95% of its 150 residents died.

Protective sequestration, the West and Overseas

When Western doctors tried to transport Eskimos to hospital, they saw them as houses of death, with some committing suicide. On November 7, the governor banned Eskimo gatherings, but many refused to obey. A teacher reported 750 deaths in just three towns where the flu would have killed 85%, although a quarter of the deaths would have been from frostbite.

In Canada, Camp Hill Hospital in Halifax admitted 249 flu cases in the first two weeks of October, with a daily average of 17.78 admissions. At the St-Jean military hospital (near Montreal), there were 245 flu patients from September 5 to December 4, with a daily average of 2.7 patients and a day with 32 admissions. Unlike what happened from winter to summer, when there were no deaths, 21 patients died (9% of those admitted).

On the same October 14 that the pandemic appeared in the capital of Alaska, in Montreal there were already 165 deaths out of 378 cases of pandemic flu. Three days later, there were another 6,283 cases and 839 deaths. The city council closed the meeting places, etc. As of November 7, 17,252 severe cases of influenza had accumulated (an estimated 100,000 people became ill) and 3,028 deaths.

In the entire province of Québec, from July to November 1918, 530,704 cases were registered. In remote Labrador villages, as in Alaska, the weakened inhabitants were unable to bury the dead, and dogs ate them. In New Brunswick there were 1,394 deaths.

In Ontario, 300,000 cases of influenza were registered, affecting mainly Toronto, Hamilton, St. Catharines, London and Conllingwood. Of the 61,063 soldiers present in Canada, 10,506 had fallen ill with the flu as of mid-December, to which had to be added 45,960 stationed outside the country. After January 1919, the epidemic subsided, as summarized by George C. Kohn (p. 60).

In Venezuela, the flu arrived in Ciudad Bolívar on October 1, on a ship from the British colony of Trinidad, where the epidemic had been declared days before. On October 16, 1918, the first case of influenza occurred in the port of La Guaira, and two days later in Caracas. On October 25, the dead began to appear lying in the streets of the poor neighborhoods of the capital. Ali

Santiago Mata: 1917 American Pandemic

Gómez, son of the president-dictator, fell ill and died without his father visiting him, for fear of contagion. On November 29, the epidemic was declared extinct in the port of La Guaira. In February 1919 the last cases were registered in Muchuquíes. According to Luis Heraclio Medina (article published on October 10, 2016) in four months 60,000 Venezuelans died.

In Argentina, the pandemic entered in October through the port of Buenos Aires: on the 16th La Nación published the first news, with the headline "The grippe should not alarm, its benign presentation." This was the case, according to Carbonetti, until the end of 1918, affecting only the coastline and the central region, with 2,237 deaths, multiplying by seven the 319 of 1917. In a second wave in May 1919, the death toll, with 12,750, would multiply by 5.7 that of the 1918 wave and by 40 that of the seasonal flu of 1917. Between the two waves, the deaths in Argentina were 14,997, this data being collected only by the "provinces", so Carbonetti assumes that, including " national territories "for which there is no data, the number of deaths would have to be multiplied by 2.4, to exceed 36,200.

At the end of October, the pandemic would enter central Chile, particularly Santiago, according to the press from Argentina and the port of Valparaíso. As documented by Marcelo López and Miriam Beltrán, the declared deaths were 6,026 in 1918 (more than double the 2,798 in 1917), 23,789 in 1919 (quadrupling the previous year's figure), and still 6,298 in 1920 and 7,228 in 1921. In total, 43,341 between 1918 and 1921.

With much more violence, according to Goulart, the pandemic hit Brazil, which had declared war on Germany over submarine attacks; since only in Rio de Janeiro —with 910,710 inhabitants— it went from 48 deaths from influenza in September 1918 to 15,000 deaths (estimating that almost two thirds of the population sickened), being the case that in a single day, the 22 October, 930 people died of flu out of a total of 1,073 deaths in the city (87%).

The first Brazilians to suffer the flu were the members of the medical mission sent to Dakar, who were contagious on board the ship *La Plata*, 156 of them dying according to the news sent on September 22 by the head of the mis-

Protective sequestration, the West and Overseas

sion, Nabuco Gouvêa. Meanwhile, in the port of Rio de Janeiro no quarantines were applied, and there were no means with which to disinfect the ships, so the flu arrived on September 14 aboard the English ship *Demerara*, which had stopped from Liverpool in Lisbon, Recife and Salvador. Although the ship's patients were taken to the Hospital de Isodamientos, the rest of the passengers went to the city and several became ill in the following days.

Paradoxically, the Rio de Janeiro press had openly affirmed the hoax that the flu was a German invention spread through submarines, for example in the newspaper *A Careta* of October 5, 1918, with the drawing *"Bacilomarino.* Another ally for the central empires ", and in words:

> The Spanish influenza and the dangers of contagion. This disease is a creation of the Germans who spread it all over the world through their submarines. Our officers, sailors and doctors of our squad, who left a month ago, pass through the hospitals on the front, catching it on the way and being victims of the treacherous bacteriological creation of the Germans, because in our opinion the mysterious disease was manufactured in Germany , charged with virulence by the German wise men, bottled and then distributed by the submarines that are in charge of spreading the carafes near the coasts of the allied countries, so that, carried by the waves towards the beaches, the bottles, collected by people innocent, spread the terrible disease throughout the universe, thus forcing the neutrals to remain neutral.

In Peru, according to the study led by G. Chowell, after the initial first wave of relatively benign pandemic influenza registered in Lima between July and September, "a severe and synchronized pandemic wave" appeared in three cities studied, between November 1918 and February 1919 (as opposed to the usual claim, which can still be found today on websites and government documents of Peru, that the flu did not affect the country until March 1919), plus a third "severe pandemic wave" that in Lima lasted from January to

Santiago Mata: 1917 American Pandemic

March 1920 and in Ica from July to October 1920 (this third wave did not occur in Iquitos).

The estimate of flu victims is made in this study based on the "accumulated excess in mortality rates", which would have been 1.6% in Lima and 2.9% in Iquitos for the whole waves of pandemic flu. By comparison, this excess would have been 0.4% in Denmark, 0.7% in Mexico City, 1.1% in Europe, 1.9% in Toluca (Mexico) and 4.9% in India.

November brought the pandemic to some remote places like German Samoa, in that case by sea from New Zealand. On November 4, 1918, the ship *Talune* arrived in Apia, having sailed from Auckland with the flu and after been quarantined in Fiji. In Samoa, the port attendant was not aware of an epidemic in Auckland and let the passengers off, including six serious cases of flu. Within a week, the pandemic had spread to the main island of Upolu and neighboring Savai'i.

The epidemic spread uncontrollably and 8,500 people died, representing 22% of the population (this being the same percentage of deaths among adult women, compared to 30% in adult men and 10% in children). In 1947, the UN described this episode as "one of the most disastrous epidemics recorded in the world during this century, in terms of the proportion of deaths in the population."

American Samoa was not affected by the flu, thanks to the blockade organized by Governor John Martin Poyer, and the same happened in the French colony of New Caledonia. In contrast, in Nauru 16% of the population died, in Tonga 8% and in Fiji 5% (9,000 people).

The survivors of ex-German Samoa accused the New Zealand administrator (as the military of that country occupied it in 1914), Lieutenant Colonel Robert Logan, for failing to quarantine the *Talune* and refusing medical aid offered from American Samoa. A royal commission found evidence of administrative carelessness and insufficient judgment.

Protective sequestration, the West and Overseas

Logan left Samoa in early 1919, and in his August 8 report he called the Samoans' upset as "temporary because as children they will forget it as long as they are treated with care; later they will remember everything that was done for them in the previous four years." He was replaced by a new administrator, Colonel R.W. Tate from 1920 to 1923. In 2002, New Zealand Prime Minister Helen Clark officially apologized to the Samoan people for the actions of the New Zealand authorities.

Santiago Mata: 1917 American Pandemic

The third wave

In the winter following the signing of the armistice in Europe, the third wave of pandemic influenza took place, the least lethal, except for regions that had not been reached by the two previous waves, as was the case in Australia.

The pandemic reached Sydney in January 1919 and the official Australian count would be 12,000 fatalities. The Queensland government closed its borders and established quarantine camps in the south where all travelers who wanted to enter the state had to stay for seven days. The two largest fields were those of Wallangarra and Coolangatta. Despite these measures and the careful examination of the ships, the disease entered that state on May 3, 1919, diagnosed in washerwomen at the Kangaroo Point Hospital in Brisbane. From there, any attempt to contain it was useless.

In Brisbane there were 9,570 flu cases and in other parts of Queensland 11,099. The following winter (1919-1920) there was less flu in the city (1,483 cases) but more in the rest of the state (17,319). As in the rest of the pandemic cases, most of the 830 deaths in Queensland in 1919 were young. 69 of the 596 inhabitants of the Barambah Aboriginal Reserve (11.6% of the population) also died.

If in any case it was proven that the low immunity acquired by non-European races made them easy targets for this pandemic, it was that of the island of Réunion, a French colony in the Indian Ocean that had been saved from the epidemic until the ship *Madonna* arrived from Marseille in March 1919.

Although military law was not applied in Réunion, the island would pay dearly for having provided 14,423 volunteers to defend the mother country. 949 of these soldiers died on the field of honor, among them the famous aviator Roland Garros.

But death had not taken its final toll when on March 31, 1919, the *Madonna* returned 1,603 soldiers to the island. The longshoremen and prisoners who unloaded the ship and cleaned the cove were the first to be affected by flu accompanied by symptoms of meningitis-encephalitis, which soon led to their death. After having landed the virus, the *Madonna* set sail on April 13 for metropolitan France, loaded with sugar.

Santiago Mata: 1917 American Pandemic

It was assumed that the virus traveled on the sand used as ballast for the ship, which some deduced may have been used to bury people with the pandemic. The truth is that some witnesses, such as Adolphe Saint-Louis, declared that the sand gave off a foul smell, was wet and had bones. This witness fell ill immediately after working on the ship, for four days, until April 3.

Infected or not by that sand, the ship was carrying sick soldiers, such as the surnamed Annette, who died the same afternoon of the landing, 500 meters before reaching his home. On April 3, as had happened in so many other places, the Réunion newspapers published notices about the presence of "hitherto benign" flu, the examination of which "does not allow us to speak of the Spanish flu, but of simple flu." But that same day Dr. Archambeaud published in *Le Progrès* that it was "a flu with amazing contagious powers, since in 24 hours it has affected 80 detainees and officials of the central prison."

Still on April 9 the *Nouveau Journal* spoke of 77 patients "with fever, dysentery." The next day, in *Le Progrès*, the head of the colony's health service, Dr. Jules Auber, declared: "It is a fast contagion flu epidemic, but it is not the Spanish flu. It's a simple flu epidemic ".

By then it had already been questioned that the pandemic flu was caused by the Pfeiffer bacillus. In 1892 the first virus had been discovered —that of the tobacco mosaic— and the "filtering virus" found at the Pasteur Institute by René Dujarric de la Rivière during the second wave of the pandemic was considered responsible for the disease. The opinion was shared by authors such as the French Nicolle Lebailly, the Germans Selter and Krtise, and the Italian Pontano.

But on the island of Réunion the doctors still did not agree on what disease they were treating. On April 14, Dr. Victor Mac-Auliffe claimed that it was Spanish flu, and Dr. Brochard, secretary general of the island's government, denied it. No quarantine measures were taken, and the capital-port, Saint-Denis, was devastated by the pandemic: there were 20 deaths on April 12 and on April 16 there were 62. Even on that day an official statement stated that there had been "no case in the colony of the flu called Spanish ". On

The third wave

Easter week 1919 there were almost a thousand deaths in Saint-Denis.

Those responsible for caring for the population abandoned their posts, many due to illness, and rice was lacking, increasing the famine and deaths. On May 5, 94 people died in Saint-Denis. Just over 40 km to the west, in the non-coastal city of Saint-Paul, Dr. Gabriel Martin, mayor and only doctor in that city, recorded between 150 and 200 deaths a day. According to him, there were two types of flu: the one that caused bleeding from the nose, from which it was possible to cure, and the fulminant. Those who suffered from this "walked normally, then stumbled, leaned against a wall, and fell dead."

The *Nouveau Journal* lamented on April 18 that the deaths in the capital-port, which were 12 a month before, now were between 8 and 10 a day ... And that was the last issue of the newspaper, as all journalists fell ill. The island had not been supplied with drugs throughout the war, and it was necessary to wait for the *Queen Mary* sailing ship to carry 42 kilos of quinine and 12 kilos of aspirin from Mauritius on April 19.

In the absence of journalists and administrators, the victims could no longer be counted. By fleeing the capital, the inhabitants of Saint-Denis brought the epidemic to the rest of the island. Only in the high and isolated places were people safe. As reported by Gaüzère and Aubry, the religious Daughters of Mary and the members of the Conferences of Saint Vincent de Paul who went out of their way to help the people reported macabre pictures:

> Between 180 and 200 corpses lying on the ground, voracious hungry dogs or pigs stealthily tear off an arm, a piece of meat, and flee the city with their terrible theft. If the door of these silent houses is opened, a swarm of dirty flies buzzes out and the terrified eyes discover in the beds of the humble country house, two, three, four rigid bodies that decomposition makes unrecognizable.

Of the mass graves dug then by prisoners, there are still five visible today. Even the priest of the cathedral, Father Thoué, died after blessing the corpse

Santiago Mata: 1917 American Pandemic

of a Chinese man named Chaming. A new cemetery had to be opened. The decline of Saint-Denis, manifested because it went from 37,825 inhabitants in 1861 to 27,392 in 1902, was consummated by counting 21,538 inhabitants in 1921.

The epidemic subsided at the beginning of May, according to local legend thanks to the help of heaven that formed on May 11 a small cyclone that in just over an hour would have cleaned the island of all impurity. However, there were still dozens of deaths afterwards. As elsewhere, doctors reported that the pandemic hit those under 40 years of age, with an estimated 2,000 deaths in the capital (over 25,000 inhabitants, 544 in Saint-André (11,000 inhabitants), 600 in Saint-Benoît (10,500 inhabitants), 868 in Saint-Paul (18,000 inhabitants) and 312 in the port (4,000 inhabitants). On the mountain, the dead were few: 28 in Salazie and Hell-Bourg, 35 in Plaine-des-Palmistes, 27 in Entre-deux.

Against this official data, which would not exceed 4,500 deaths, Gaüzère and Aubry's estimates them between 7,000 and 20,000, or between 4 and 11% of the total population of 175,000 people. These same authors indicate that the flu affected one billion people, half of the then 1.86 billion inhabitants of the Earth, causing the death of 50 million.

In mainland France, fatalities were estimated at 408,000 people. According to Erkoreka, the mortality that in Madrid had been 1.31 per thousand inhabitants between May and June, rose in October to 5.27 per thousand, while in Paris it was estimated at 6.08 per thousand and in south-western Europe between 10.6 and 12.1 per thousand, mainly affecting people between 15 and 44 years of age, a group in which pandemic influenza represented 68.2% (Paris) and 66.3% (Madrid) of deaths, while the non-pandemic flu of 1916, 1917 and 1921 accounted for 19.7%, 12.5% and 21% of the deaths in that age group.

In Italy, Alfani and Melegaro place between 300,000 and 400,000 fatalities, having suffered the disease between five and seven million, because although the Istituto Centrale di Statistica evaluated the deaths at 274,051 (in 1918 in the Italian Peninsula), it did not include deaths due to pneumonia or

The third wave

bronchopneumonia and already in 1925 Giorgio Mortara estimated the deaths at 600,000. The estimates of Patterson and Pyle (325 to 350,000) or Niall Johnson (390,000) seem more appropriate to these authors.

For some countries like Poland, it was still impossible to assess the mortality caused by the 1918 flu in 2016, according to Jan Wnęk. The same can be said for a number of central European countries and Soviet Russia.

The restrictions on maritime traffic imposed in Japan appear to have caused the pandemic to be less severe there than elsewhere in Asia (257,363 officially killed as of July 1919, representing a mortality rate of 4.25 per thousand).

Santiago Mata: 1917 American Pandemic

The replicants won over winners and losers

How could a disease as old as the flu cause the greatest mortality in history at a time when scientific advances were supposed to stop any pandemic? The answer, possibly, is that in fact the means to prevent the catastrophic effects of the H1N1 virus already existed then, but these measures were not taken. Consciously or unconsciously?

Although the country and the date where it arose is known, after a century there are still great unknowns about the origin and impact of that deadly disease. Genetic reconstruction has made possible to verify that the eight genes that make up the H1N1 virus come from birds, but at the same time that the virus as such was more different from the viruses found in North American birds of that time, than those viruses of birds from 1917-1918 preserved in laboratories are with respect to viruses found in current birds.

Rather than coming from a rare bird, the H1N1 of the 1917-1918 pandemic was such a rare avian virus that it only had the essentials, the genes, of birds, so that those who have rebuilt it do not explain where it could have arisen and almost seem to suggest that the compound was the work of extraterrestrial forces.

Equally inexplicable seems the reaction it provoked, perhaps because its novelty disproportionately stimulated the immune defenses, with the result that those who, due to weakness - too young or too old - had little means to fight the virus, survived and, on the other hand, strong organisms, with a disproportionate immune alarm that triggered the congestion of the lungs - the so-called cytokine storm -, caused more damage to themselves, resulting in this reaction more harmful than the virus.

It is not difficult —but it is paradoxical— to see in this disease one more case in which the human being, the stronger he pretends to be, in reality becomes more fragile. The men of that time, relied on their own intelligence and willpower to prevail over others with war, but were mortally trapped in networks from which they seemed unable to escape.

Just as the Great War was paradoxically promoted as the war that was to end all wars, a progressive-pacifist government ended up dragging the American

Santiago Mata: 1917 American Pandemic

people into war, selling them bonds that could possibly be called Liberty Bonds, but without doubt not health bonds.

In a proud age, when men had believed themselves to be omnipotent, this economic boost was supposed to save human lives. But many people who paid to get rid of Central European authoritarianism -whether real or imagined the threat it posed-, and many others who had nothing to do with the war, paid with their lives for the incapacity of the most intelligent and strong men to fight a virus. And the victims of a being so small that it cannot even be called an organism - the closest thing to life we can preach from viruses is that they are replicants - possibly reached the hundred million, tripling or even tenfold the victims of the First World War (depending on whether we accept that this war caused 10 million deaths or up to 31).

Paradoxically, those who broke the sanitary regulations and facilitated the development of the deadliest pandemic that humanity has known, did so in order not to stop the largest propaganda campaign in history: the US Liberty Bonds, which would provide their government with 17,000 million dollars, a figure never imagined in the service of the war, but not of the health of the people whose lives were risked by bringing them together to ask for money or, even more, to train them as soldiers.

The spread of the epidemic, the Liberty Bonds campaign and the training of millions of young people as improvised warriors, derived from the way a government and a group of politicians who believed to be deeply pacifists wanted to conduct the war. The largest military mobilization - at least up to that moment the fastest and for its proportionally larger volume - in history was made apparently combining the spontaneity and the security of having the best guarantees regarding its financing and health.

For the first time it was tried to avoid that, in a war, the diseases and not the bullets, were the main causes of the deaths, and the attempt ended with a resounding failure. The naivety of the pacifists who believed they could confront the Prussian military tradition with an improvised army, although certainly numerous, had to do, more than with the fearsome German guns, with tiny viruses whose existence was not even known.

The replicants won over winners and losers

It seems as if nature made fun of the pretense of planning a victory to the millimeter, but not only on the part of the Allies, whose plans it should have come in 1919. As if the prediction that three little shepherds heard on July 17, 1917 in the small Portuguese town of Fatima, according to which "the war will soon end", became a reality. The victory was advanced, ruining the German plans, set in the offensive of March 21, 1918, but also surpassing the plans of the Allies, who could not attribute the victory to their intelligent planning, their money, or the strength of their weapons.

The US government did not act against the deadly pandemic flu that emerged in its country and was exported by its soldiers until it had an alibi to distract another culprit. When the Spanish press spoke of the epidemic, it was named the Spanish flu. An outrageous name that brings the evils suffered by half humanity - by all of humanity at that time - on a country that was not even at war. But, as absurd as it may have been, the theory that the flu emerged in Spain caught on, and lasts until today.

The first conclusion, a century after that pandemic, for a minimum of coherence and respect, should be not to allow that disease to be called Spanish flu, at least not without clarifying that it arose in the United States in 1917-1918.

The administration whose army spread the disease to the whole world after abandoning its —at least relative— neutrality in the conflict, would end up hanging the stigma of guilt in its origin to a neutral country —Spain—, and the government that refused to investing funds and efforts in fighting the pandemic so as not to stop taking money from its citizens and turning them into military personnel, would blame civilians for the spread of the epidemic, instead of acknowledging that it had arisen in military camps.

Even after recognizing that the flu that would ravage the world in 1918 emerged in the United States, its politicians preferred to accept the naive option that a pandemic that requires large concentrations of people to spread, would have arisen in the Kansas desert - and in places about to be abandoned, as is the town of Santa Fe in Haskell County - delaying its start date, rather than admitting evidence that the flu had emerged, as it always

Santiago Mata: 1917 American Pandemic

does, in the fall, that it became in a deadly epidemic in military camps in the last months of 1917, and that it spread to Europe after the transfer of the 32nd Infantry Division and the rest of the US Army.

Comparison with the covid-19 pandemic

When the Spanish version of this work was published in 1917, the covid-19 pandemic had not started, and when its English version is published in 2020, it has not ended, so any comparison between the two pandemics will necessarily be limited. However, I think that, within these limits, two considerations can be made, relative to the moment of their beginning and to the human reaction to them.

The pandemic declared in 1918 was the consequence of an epidemic disease that began in 1917, in the same way that the pandemic declared in 2020 was the consequence of an epidemic disease that began in 2019, which is precisely why it is called covid-19.

The second statement can be proven with reliable data, while there is less data on the first. Some of them have been exposed in this work.

In both pandemics, the difference between their dates of origin and the time when each pandemic was declared -although in 1918 there was no universally accepted mechanism to give that qualification to an infectious disease-, opens the search for an explanation.

In the case of the pandemic that broke out in 1917 in the United States, the medical measures to fight it collided head-on with the need to bring together large numbers of people to pay the Liberty Bonds, with which to pay and predictably win the war , and of bringing together large numbers of soldiers - precisely those among whom the pandemic flu emerged - to send them to the European front.

The fact that the war effort prevented the sanitary effort does not in itself prove that the second was not carried out for that reason. At least there is no evidence for the months in 1917 when the epidemic broke out. The known demands of doctors about the suspension of military transport are from September 1918, therefore when the pandemic was already a fully extended reality. Still, the obligation of the US rulers to have known the situation and taken action remains real and open to scrutiny.

Santiago Mata: 1917 American Pandemic

In the case of the epidemic that emerged in 2019 and was declared a pandemic in 2020, there are also enormous economic interests faced with the measures that could or can be taken against its extension. It is clear that the government of the People's Republic of China also avoided providing timely information on the severity and extent of the epidemic.

As for the rest of the governments, it is obvious that there were those who also preferred to turn a deaf ear to the damage to health in order not to damage their economic and political projects, but this is not the place to evaluate them and even less to judge them.

The pandemic that emerged in 1917 spread without restraint and claimed many more lives than, at the moment, the one that emerged in China in 2019. However, in 1917 there were far fewer means to know and avoid the disease. This relative ignorance, no doubt, made it apparently more difficult to foresee the consequences of inaction or even - in terms of gathering people - of action that was directly harmful in epidemic terms.

The responsibility of politicians, as can be seen in both cases, can be immense, and it is certainly a necessary subject for reflection, as well as unattainable from the limits of this work.

Bibliography

Guido **Alfani** and Alessia **Melegaro**: *Pandemie d'Italia. Dalla peste nera all'influenza suina: l'impatto sulla società*. EGEA, Milan, 2010, 224 pages.

Raquel **Almansa**: "El virus gripal pandémico H1N1", in *Epidemiología Molecular de Enfermedades Infecciosas*, October 10, 2012 (http://epidemiologia-molecular.com/virus-gripal-pandemico-h1n1/).

C. E. **Ammon**: "Spanish flu epidemic in 1918 in Geneva, Switzerland". Eurosurveillance Volume 7, Issue 12, 01/Dec/2002 (https://www.eurosurveillance.org/content/10.2807/esm.07.12.00391-en?crawler=true#html_fulltext).

Army War College, *Historical section: The Aisne and Montdidier-Noyon operations*. Washington, 1922, Government Printing Office, 52 pages. (https://archive.org/details/aisnemontdidiern01unit)

G. **Audeoud**: "La gripe en 1918 dans la 1re division". *Revue Militaire Suisse*, vol 68, 1923, cahier 2.

Leonard P. **Ayres**: *The War with Germany. A Statistical Summary*. Washington, 1919, Government Printing Office, 154 pages. (http://www.gwpda.org/docs/statistics/statstc.htm)

John M. **Barry**: *The Great Influenza. The Epic Story of the Deadliest Plague in History*. Penguin, 2005, 546 pages.

"The site of origin of the 1918 influenza pandemic and its public health implications". *Journal of Translational Medicine*, January 20, 2004. (https://www.ncbi.nlm.nih.gov/pmc/articles/PMC340389/)

Frieder **Bauer**: *Die Spanische Grippe in der deutschen Armee 1918: Verlauf und Reaktionen*. Dissertation for the degree of Doctor of Medicine at the Heinrich Heine University of Düsseldorf, 2014, 142 pages. (https://docserv.uni-duesseldorf.de/servlets/DerivateServlet/Derivate-38595/DissertationFBauerPDFA.pdf)

Robert B. **Belshe**: "The Origins of Pandemic Influenza — Lessons from the 1918 Virus". *The New Englang Journal of Medicine*, 2005; 353:2209-2211,

Santiago Mata: 1917 American Pandemic

November 24, 2005:

(http://www.nejm.org/doi/full/10.1056/NEJMp058281)

Laurel Downing **Bill**: "1918: The big sickness spreads across Alaska". *Senior Voice*, April 1, 2015, Vol. 38, N. 4:

(http://www.seniorvoicealaska.com/story/2015/04/01/columns/1918-the-big-sickness-spreads-across-alaska/732.html)

Debra E. **Blakely**: *Mass Mediated Disease: A Case Study Analysis of Three Flu Pandemics and Public Health Policy*. Lexington Books, Oxford, 2006, 180 pages.

Wyndham Bolling **Blanton** and Ernest Edward **Irons**: "A Recent Epidemic of Acute Respiratory Infection at Camp Custer, Mich. Preliminary Laboratory Report". *The Journal of the American Medical Association*, December 14, 1918, vol. 71, pages 1988-1191. Published as a 12-page supplement, NLM id 101501130. (http://resource.nlm.nih.gov/101501130)

L.S. **Block**: "The Invisible Enemy of 1918". *Yankee Magazine*, January-February 1982, Dublin, New Hampshire:

(https://newengland.com/yankee-magazine/living/new-england-history/the-influenza-pandemic-of-1918/)

Rupert **Blue**: *Report of the Surgeon General U.S. Army to the Secretary of War 1919*. Vol. I, Washington, Government Printing Office, 1919, 1.364 pages. (https://archive.org/details/reportofsurgeong19191unit)

Dr. Alfred Jay **Bollet**: *Plagues & Poxes: The Impact of Human History on Epidemic Disease*. Demos Medical Publishing, New York, 2004, 480 pages.

Françoise **Bouron**: "La grippe espagnole (1918-1919) dans les journaux français", *Guerres mondiales et conflits contemporains*, 2009/1 (n. 233), pages 83-91:

(https://www.cairn.info/revue-guerres-mondiales-et-conflits-

Bibliography

contemporains-2009-1-page-83.htm)

Nancy **Bristow**: *American Pandemic: The Lost Worlds of the 1918 Influenza Epidemic*. Oxford University Press, Oxford - New York, 2012, 296 pages.

Carol R. **Byerly**: *Fever of War: The Influenza Epidemic in the U.S. Army During World War I*. NYU Press, New York - London, 2005, 251 pages.

"The U.S. Military and the Influenza Pandemic of 1918–1919", *Public Health Reports*, 2010, n. 125 (supplement 3), pages 82-91:

(https://www.ncbi.nlm.nih.gov/pmc/articles/PMC2862337/)

Adrián **Carbonetti**: "Historia de una epidemia olvidada. La pandemia de gripe española en la Argentina, 1918–1919". *Desacatos* , n. 32, Mexico, January-April 2010:

(http://www.scielo.org.mx/scielo.php?script=sci_arttext&pid=S1405-92742010000100012)

Siddharth **Chandra**, Eva **Kassens-Noor**: "The evolution of pandemic influenza: evidence from India, 1918–19". *BMC Infectious Diseases*, 2014; vol . 14: 510. (https://www.ncbi.nlm.nih.gov/pmc/articles/PMC4262128/)

K.F. **Cheng**, P.C. **Leung**: "What happened in China during the 1918 influenza pandemic?". *International Journal of Infectious Diseases*, Volume 11, n. 4, July 2007, pages 360–364.

(http://www.ijidonline.com/article/S1201-9712(07)00035-5/fulltext)

Gerardo **Chowell** (dir.): "The 1918–1920 influenza pandemic in Peru". *Vaccine*, July 22, 2011; 29 (Suppl 2):

B21–B26. doi: 10.1016/j.vaccine.2011.02.048.

(https://www.ncbi.nlm.nih.gov/pmc/articles/PMC3144394/)

"Mortality patterns associated with the 1918 influenza pandemic in Mexico: evidence for a spring herald wave and lack of pre-existing immunity in older

Santiago Mata: 1917 American Pandemic

populations". *Journal of Infectious Diseases*, August 15, 2010; 202(4): 567–575. doi: 10.1086/654897.

(https://www.ncbi.nlm.nih.gov/pmc/articles/PMC2945372/)

Anne **Cipriano Venzon**: *The United States in the First World War: An Encyclopedia*. Routledge, New York - London, 2012, 850 pages.

George B. **Clark**: *The American Expeditionary Force in World War I: A Statistical History, 1917–1919*. McFarland, Jefferson, North Carolina, 2013, 368 pages.

Edward M. **Coffman**: "Peyton C. March: Greatest Unsung American General of World War I". *Military History Quarterly*, summer 2006. (http://www.historynet.com/peyton-c-march-greatest-unsung-american-general-of-world-war-i.htm)

Richard **Collier**: *The Plague of the Spanish Lady: the influenza pandemic of 1918-1919*. Atheneum, Richmond, Texas, 1974, 376 pages.

Douglas B. **Craig**: *Progressives at War: William G. McAdoo and Newton D. Baker, 1863–1941*. JHU Press, Baltimore, Mariland, 2013, 552 pages.

Alfred W. **Crosby**: *Epidemic and Peace: 1918*. Greenwood Press, Westport, Connecticut, 1976, 337 pages.

Josephus **Daniels**: (Annual) *Report of the Secretary of the Navy. Miscellaneous Reports*. 1919.

Pierre **Darmon**: "Une tragédie dans la tragédie : la grippe espagnole en France (avril 1918-avril 1919)". In: *Annales de démographie historique*, 2000-2, pages 153-175.

Beatriz **Echeverri Dávila**: *La Gripe Española. La pandemia de 1918-1919*. Centro de Investigaciones Sociológicas, Siglo XXI de Editores, Madrid, 1993, 194 pages.

Herman **Elwyn**: "Pneumonia at Camp Greene: A Few Considerations from a

Bibliography

Clinical Standpoint". *Southern Medical Journal*, Birmingham, Alabama, December 1918, n. 12, pages 780-785.

Antón **Erkoreka**: "The Spanish influenza pandemic in occidental Europe (1918–1920) and victim age". Influenza and Other Respiratory Viruses, 4: 81–89. doi:10.1111/j.1750-2659.2009.00125.x

Luis **Español**: *Leyendas negras. Vida y obra de Julián Juderías*. Junta de Castilla y León, Salamanca, 2007, 413 pages.

Erich von **Falkenhayn**: *Die oberste Heeresleitung, 1914-1916 in ihren wichtigsten Entschliessungen*. Mittler und Sohn, Berlin 1920, 284 pages (University of Toronto: https://archive.org/details/dieobersteheeres00falk).

Wilton B. **Fowler**: *British-American Relations 1917-1918: The Role of Sir William Wiseman. Supplementary Volume to The Papers of Woodrow Wilson.* Princeton University Press, New Jersey, 1969, 350 pages.

Alfred **Friedlander** y otros: "The Epidemic of Influenza at Camp Sherman, Ohio", *Journal of the American Medical Association*, November 16, 1918, pages 1652-1656.

Bernard **Gaüzère** y P. **Aubry**: "La pandémie de grippe espagnole de 1918-1919 á la Réunion". *Médecine et Santé Tropicales*, 2015 ; 25 : 13-20. doi : 10.1684/mst.2014.0408.

Adriana da Costa **Goulart**: GOULART, Adriana da Costa. "Revisitando a espanhola: a gripe pandêmica de 1918 no Rio de Janeiro". *História, Ciências, Saúde-Manguinhos*, January-April 2005, vol. 12, n. 1, pages 101-142.

Marc **Hieronimus**: *Krankheit und Tod 1918: zum Umgang mit der Spanischen Grippe in Frankreich, England und in dem Deutschen Reich*. LIT Verlag Münster, 2006, 220 pages.

James E. **Higgins**: *Keystone of an Epidemic: Pennsylvania's Urban Experience During the 1918--1920 Influenza Epidemic*. April 22, 2009. Thesis for obtaining the degree of Doctor of Philosophy in the Department of History, Lehigh University, ProQuest LLC, 2009, Ann Arbor, Michigan, 270 pages.

Santiago Mata: 1917 American Pandemic

Brian L. **Hoffman**: "Influenza activity in Saint Joseph, Missouri 1910-1923: Evidence for an early wave of the 1918 pandemic". November 17, 2011, *PLOS Currents Influenza*, November 22, 2011:

(https://www.ncbi.nlm.nih.gov/pmc/articles/PMC3221054/)

Mark **Honigsbaum**: *Living with Enza: The Forgotten Story of Britain and the Great Flu Pandemic of 1918*. Macmillan, London, 2009, 237 pages.

Tony **Hopfinger**: "Part 3: How the Alaska Eskimo village Wales was never the same after 1918 flu". *Anchorage Daily News*, May 27, 2012. Updated: September 27, 2016.

Niall **Johnson**: *Britain and the 1918-1919 Influenza Pandemic: A Dark Epilogue*. Routledge, 2006, 288 pages.

Bogumiła **Kempińska-Mirosławska** and Agnieszka **Woźniak-Kosek**: "The influenza epidemic of 1889-90 in selected European cities--a picture based on the reports of two Poznań daily newspapers from the second half of the nineteenth century". *Medical Science Monitor*, December 10, 2013; 19:1131-41. Doi: 10.12659/MSM.889469:

(https://www.ncbi.nlm.nih.gov/pmc/articles/PMC3867475/)

David **Killingray**, Howard **Phillips**: *The Spanish Influenza Pandemic of 1918-1919: New Perspectives*. Routledge, London - New York, 2003, 384 pages.

George Childs **Kohn**: *Encyclopedia of Plague and Pestilence: From Ancient Times to the Present*. Infobase Publishing, New York, 1995, 529 pages.

Olivier **Lahaie**: "L'épidémie de grippe dite «espagnole» et sa perception par l'armée française (1918-1919)". P. 102-109 of *Revue Historique des Armées*. 262, 2011.

Marcelo **López** y Miriam **Beltrán**: "Chile entre pandemias: la influenza de 1918, globalización y la nueva medicina". *Revista chilena de infectología*. Vol. 30, n. 2, Santiago, April 2013.

Bibliography

(http://dx.doi.org/10.4067/S0716-10182013000200012)

Charles **Lynch** (editor): *The Medical Department of the United States Army in the World War. Volume I.* The Surgeon General's Office. Government Printing Office, Washington, 1923, 1.389 pages.

Volume VI. Sanitation. Government Printing Office, Washington, 1926, 1141 pages.

Volume IX: Communicable and Other Diseases. Washington, U.S. Government Printing Office, 1928:

(http://www.ibiblio.org/hyperwar/AMH/XX/WWI/Army/Medical/IX/)

Volume XII. Pathology of the Acute Respiratory Diseases and of Gas Gangrene following War Wounds. Prepared by George R. Callender and James F. Coupal, Washington, 1929, 577 pages.

Irving P. **Lyon** (dir): "Some Clinical Observations on The Influenza Epidemic at Camp Upton". *Journal of the American Medical Association*, June 14, 1919, pages 1726 – 1731.

Howard **Markel** (dir): *A Historical Assessment of Nonpharmaceutical Disease Containment Strategies Employed by Selected U.S. Communities During the Second Wave of the 1918-1920 Influenza Pandemic.* The University of Michigan Medical School Center for the History of Medicine, January 31, 2006, 275 pages:

(http://chm.med.umich.edu/wp-content/uploads/sites/20/2015/01/DTRA-Final-Influenza-Report.pdf)

William G. **McAdoo**, W.E. **Woodward**: *Crowded Years: The Reminiscences of William G. McAdoo.* Houghton Mifflin, 1931.

Carey P. **McCord**: "The Purple Death: Some Things Remembered about The Influenza Epidemic of 1918 at One Army Camp". *Journal of Occupational Medicine*, November 1966, Vol. 8 , n. 11, pages 593-598.

Santiago Mata: 1917 American Pandemic

Leon S. **Medalia**: "Influenza Epidemic at Camp MacArthur: Etiology, Bacteriology, Pathology, and Specific Therapy". *Boston Medical and Surgical Journal*, March 20, 1919; n. 180, p. 323-330, DOI: 10.1056/NEJM191903201801201

Luis Heraclio **Medina Canelón**: "Octubre de 1918: La epidemia de de Gripe Española arrasa con Venezuela". October 10, 2016:

https://www.facebook.com/notes/luis-heraclio-medina-canelon/octubre-de-1918-la-epidemia-de-gripe-espa%C3%B1ola-arrasa-con-vene-zuela/10209447181592895/

Miriam Grace **Mitchell** and Edward Spaulding **Perzel**: *The echo of the bugle call: Charlotte's role in World War I*. Heritage Printers Inc., Charlotte, North Carolina, 1979, 87 pages.

Martin **Motte**: "Ferdinand Foch (1851-1929)". *Cahiers du CESAT*, n. 26, December 2011, p. 7-11.

George **Newman**: *Report on the pandemic of influenza 1918–19*. Reports on public health and medical subjects n. 4, Her Majesty's Sationery Office, London, 1920, 577 pages.

Donald R. **Olson**, Lone **Simonsen**, Paul J. **Edelson** and Stephen S. **Morse**: "Epidemiological evidence of an early wave of the 1918 influenza pandemic in New York City", in *Proceedings of the National Academy of Sciences of the United States of America*, August 2, 2005, volume 102(31): pages 11059–11063:

(https://www.ncbi.nlm.nih.gov/pmc/articles/PMC1182402/)

Sandra **Opdycke**: *The Flu Epidemic of 1918: America's Experience in the Global Health Crisis*. Routledge, London - New York, 2014, 234 pages.

Mark **Osborne Humphries**: *The Last Plague: Spanish Influenza and the Politics of Public Health in Canada*. University of Toronto Press, 2013, 348 pages.

Karl David **Patterson**: *Pandemic Influenza, 1700-1900: A Study in Historical*

Bibliography

Epidemiology. Rowan & Littlefield, Totowa, New Jersey, 1986, 118 pages.

Howard **Phillips**: *Black October: The Impact of the Spanish Influenza Epidemic of 1918 on South Africa*. Pretoria, Government Printing, 1990.

(With David Killingray.) *The Spanish Influenza Epidemic of 1918-1919: New Perspectives*. 2003, 384 pages.

"The Re-Appearing Shadow of 1918: Trends in the Historiography of the 1918-19 Influenza Pandemic," *Canadian Bulletin ofMedical History* 21,1 (2004): 121-34.

María Isabel **Porras Gallo**: *Una ciudad en crisis. La epidemia de gripe de 1918-19 en Madrid*. Faculty of Medicine of the Complutense University of Madrid (doctoral thesis), 1994, 756 pages.

Reichskriegsministerium: *Influenza. Gelbkreuz. Sanitätsbericht über das Deutsche Heer (Deutsches Feld- und Besatzungsheer) im Weltkriege 1914/1918*. Volume III: *Die Krankenbewegung bei dem Deutschen Feld- und Besatzungsheer*. E. S. Mittler & Sohn, Berlin 1934.

Geoffrey **Rice** and Linda **Bryder**: *Black November: The 1918 influenza pandemic in New Zealand*. Canterbury University Press, Christchurch, 2005, 327 pages.

Jeffrey R. **Ryan**: *Pandemic Influenza: Emergency Planning and Community Preparedness*. CRC Press, Boca Raton, Florida, 2008, 280 pages.

Lawrence D. **Schuffman**: "Capitalizing on American Pride and Patriotism. Funding of the First World War through The Liberty and Victory Loan Bonds 1917-1923". *Paper Money*, n. 259, January-February 2009, pages 3-33. (http://www.moaf.org/exhibits/checks_balances/woodrow-wilson/materials/Schuffman_SPMC_article.pdf)

Richard E. **Shope**: "The incidence of neutralizing antibodies for swine influenza virus in the sera of human beings of different ages". *Journal of Experimental Medicine*, April 30, 1936; vol. 63 (5), pages 669-684. (https://www.ncbi.nlm.nih.gov/pmc/articles/PMC2133359/)

Santiago Mata: 1917 American Pandemic

John **Shortal**: "End of Days: Responding to the Great Pandemic of 1918", p. 6-26 of James Jay **Carafano**, Richard **Weitz**: *Mismanaging Mayhem: How Washington Responds to Crisis*. Praeger Security International, Greenwood Publishing Group, Westport, Connecticut, 2008, 296 pages.

Lone **Simonsen**: "The global impact of influenza on morbidity and mortality", *Vaccine*, 17 (1999), p. 3-10.

Philip A. **St. John**: *History of the Third Infantry Division. Rock of the Marne*. Turner Publishing Company, Paducah, Kentucky, 1994, 120 pages.

Mike **Stobbe**: *Surgeon General's Warning: How Politics Crippled the Nation's Doctor*. University of California Press, Oakland, California, 2014, 394 pages.

Jennifer A. **Summers**,Nick **Wilson**, Michael G. **Baker**, G. Dennis **Shanks**: "Mortality Risk Factors for Pandemic Influenza on New Zealand Troop Ship, 1918". *Emerging Infectious Diseases*. December 2010; 16(12): 1931–1937. doi: 10.3201/eid1612.100429. PMCID: PMC3294590.

Richard **Sutch**: *Financing the Great War: A Class Tax for the Wealthy, Liberty Bonds for All*. Berkeley Economic History Laboratory Working Paper WP2015-09, September 2015, 66 pages:

(http://behl.berkeley.edu/files/2015/09/WP2015-09_Sutch.pdf)

Robert **Swanson**: *Domestic United States Military Facilities of the First World War 1917-1919*. Robert Swanson, Rapid City, South Dakota, 2000, 410 pages.

Jeffery K. **Taubenberger**, Johan V. **Hultin**, David M. **Morens**: "Discovery and characterization of the 1918 pandemic influenza virus in historical context". *HHS Author Manuscripts*, May 22, 2008. *Antivir Ther*. 2007; 12(4 Pt B): 581–591. (https://www.ncbi.nlm.nih.gov/pmc/articles/PMC2391305/)

Malte **Thießen**: *Infiziertes Europa: Seuchen im langen 20. Jahrhundert*. Walter de Gruyter GmbH & Co KG, Munich, 2014, 219 pages.

Spencer **Tucker**: *The Great War, 1914-1918*. Taylor & Francis,January 4, 2002, 272 pages.

Bibliography

Various authors (Defense Department): *American Military History: The United States Army in a global era, 1917-2003*. Vol. II. 2010, Government Printing Office, 525 pages.

Various authors: *History of the U.S.S. Leviathan Cruiser and Transport Forces United States Atlantic Fleet*. Brooklyn Eagle Book Department, 1919, 236 pages.

Manfred **Vasold**: *Die Spanische Grippe : Die Seuche und der Erste Weltkrieg*. Primus, Darmstadt, 2009, 142 pages.

Robert L. **Willett**: *Russian Sideshow: America's Undeclared War, 1918-1920*. Potomac Books, Inc., 2003, 327 pages.

Jan **Wnęk**: "The Spanish Flu epidemic in Poland (1918-1919)". 2016, *Epidemiology* (Sunnyvale) 6:240. doi:10.4172/2161-1165.1000240

Santiago Mata: 1917 American Pandemic

Acknowledgment

I am grateful to the engineer Antonio Luna for proposing to me to investigate together the origin of the 1918 pandemic and its possible influence on the outcome of the First World War. Although the North American origin of the so-called Spanish flu was known, he suggested me to study what happened in the military camps, thus opening the door for what I consider to be my main discovery: that the pandemic flu emerged in 1917 and was spread in the US Army facilities, and from them to the entire world.

The influence that the flu could have had on the outcome of the First World War is, in my opinion, much more difficult to assess. Since this is Antonio Luna's main field of interest, I have not been able to share with him a work that unites both fields of research. For this reason, while I wish him success in the course of his studies, I hereby record that thanks to his suggestion I entered the track of what seems to me a historical discovery of great importance.

Valladolid, November 7, 2020.

Made in the USA
Monee, IL
18 August 2021